MATERIALS SCIENCE AND TECHNOLOGIES

CHARCOAL

CHEMICAL PROPERTIES, PRODUCTION METHODS AND APPLICATIONS

Materials Science and Technologies

Additional books in this series can be found on Nova's website under the Series tab.

Additional e-books in this series can be found on Nova's website under the e-book tab.

MATERIALS SCIENCE AND TECHNOLOGIES

CHARCOAL

CHEMICAL PROPERTIES, PRODUCTION METHODS AND APPLICATIONS

ERIC N. TYRONE
EDITOR

New York

For permission to use material from this book please contact us:
Telephone 631-231-7269; Fax 631-231-8175
Web Site: http://www.novapublishers.com

Library of Congress Cataloging-in-Publication Data

ISBN: 978-1-62808-663-8

Published by Nova Science Publishers, Inc. † New York

CONTENTS

PREFACE

Charcoal is a light black residue consisting of carbon, and any remaining ash, obtained by removing water and other volatile constituents from animal and vegetation substances. In this book, the authors discuss the chemical properties, production methods and applications of charcoal. Topics include the abstention of charcoal from eucalyptus wood in a steel pilot scale kiln; application of carbon-supported metal catalysts in the field of hydrodechlorination; emission reduction theory and results in the development of a suitable small-scale, portable, continuous rice husk carboniser for rural regions in transitional economies; the significance of functionalization of carbon based adsorbents for water cleaning technologies; and charcoal meal as an old but valid test to measure intestinal transit in vivo.

Chapter 1 – Since the oldest ages, carbon has been a part of man's history. Since it's grinding for making the wonderful drawings of Grotte Chauvet, in the French province of Ardeche, to the breathtaking advances in the field of nanocarbon tubes and graphene, carbon has been in the center of man's scientific preoccupations. Isn't organic chemistry defined as "the chemistry of carbon"?

In the field of organic synthesis, C has been initially used as a support for various metallic catalysts. An extremely useful application in the field of environmental protection was the hydrodechlorination reaction (the replacement of a chlorine atom on a substrate by a hydrogen atom) allowing a dramatic lowering of toxicity. This is a very sensitive subject, since the most toxic compounds to man and to environment, identified by the Stockholm Convention on Persistent Organic Pollutants as the "The Dirty Dozen", are all organochlorinated species, either used or in use as pesticides (DDT, Mirex, Toxaphen, Alachlor, Chlordan, Aldrin, Dieldrin, Hexachlorcyclohexan,

Heptachlor), insulating oil (polychlorobiphenyls) or plainly toxic by-products in synthetic or pyrolitic processes (PCDDs and PCDFs).

The authors wish to present a review of recent applications of carbon-supported metal catalysts in the field of hydrodechlorination along with a shift that occurred in the past decade in scientists' perception of carbon, since there are more and more reports of various nanoforms of carbon acting solely as catalyst.

Chapter 2 – This chapter describes the process of developing a new rice husk conversion technology suitable to the local requirements of rice farmers in poor rural areas of the Philippines. The small-scale, portable, continuous carboniser was designed to enable heat extraction applications (crop drying, steam generation, etc.), while maintaining conversion effectiveness, with a special focus on the reduction of rural workplace carbon monoxide and smoke emissions. The chapter contains the technology optimization model methodology which aimed to achieve high operating temperatures for heat extraction, highest fixed carbon content of rice husk biochar, and the lowest emissions of carbon monoxide possible during the carbonisation process. A Response Surface Methodology was used to develop predictive models relating carboniser performance parameters (airflow rate, temperature, biochar quality, carbon monoxide emission data) to the carboniser parameters (chimney diameter and ignition chamber opening). The analysis of variance showed the importance of the chimney diameter relative to the ignition chamber opening in determining overall performance. The high temperatures and low CO emissions were associated with the high airflow rates, although proximate analysis of the biochar samples showed there was no influence of the relative sizes of the chimney and ignition chamber on the biochar fixed carbon content. Therefore, the selection of the optimal carboniser chimney and ignition chamber sizes were based solely on temperature and CO emissions.

Chapter 3 – The use of charcoal as an energy source has increased in developing countries in the last few decades. Charcoal production from wood has several advantages in regard to the direct use of forest biomass as fuel, among them: higher high heating value (HHV), lower moisture content and better resistance to microbiological degradation. Wood carbonization produces also condensable gases as tar, with its own value as fuel, and an aqueous phase (pyroligneous acid) that has found applications as a veterinarian product and herbicide. A lot of work has been produced recently about biomass carbonization. The results show the influence of temperature and heating rate on the process, as well as the char and condensate yields achievable.

The effect of the lignin and cellulose content was also studied. Several kinds of kilns have been developed.

In this work, the abstention of charcoal and tar from sawmill residues (*Eucalyptus grandis* wood), for energetic uses was studied. A pilot scale steel kiln (capacity 100 kg) for wood carbonization and condensable gases recovery, based in that proposed by Unrinin and Yatagai was built. Some modifications were performed in order to improve the massic yields of charcoal and condensables, the energetic yield of the process, and the control of the smoke flow. These modifications included the using of refractory bricks and ceramic fiber to increase thermal isolation, the addition of a chimney with a concentric heat exchanger as a condenser, a tar trap, a butterfly valve and a cyclone. The inner temperature at three different heights was measured and registered as a function of time. Several tests were carried out, varying the initial mass and monitoring temperature during ignition, pyrolysis, kiln cooling and unloading. Air inlet flow was strictly controlled. The whole process took about 20 h. Raw material and products (charcoal, tar) were characterized by proximate analysis, HHV determination and thermogravimetric analysis in inert and air atmosphere, in order to study the influence of the carbonization conditions on the properties of the carbonization products. The carbonization kiln produced a good quality charcoal (ash content and HHV in dry basis nearly 1 % and 30 MJ kg^{-1} respectively). Charcoal yields varied among 13 and 28 %. The energy conversion efficiency stayed between 34 and 47 %. Kiln modifications helped to overcome some troubles, such as incomplete wood carbonization. The recovery of condensate is possible, though only a low percentage of it is constituted by tar.

Chapter 4 – The characteristics of carbon can be varied according to the shape (granular, tubular, cloth etc.), porosity, specific surface etc. considering the requisites of technological applications. The properties of carbons are widely altered by the preparation method and by further functionalization and/or forming composites with other compounds (alumina, silica, or other minerals). The low price of raw material and carbon preparation technology is important in water cleaning technologies especially in developing countries. Further factors influencing the industrial use of the obtained activated (or modified) carbon are the environmentally friendly byproducts, end products as well as application processes. The carbon-based adsorbents saturated with hazardous inorganic and organic compounds should be reclaimed or recycled into the environment without any risk. The carbonization and activation method worked out in our laboratory permits the use of various raw materials resulting in the same end product. The properties of activated carbon can be easily designed by the functionalization of carbon-based low cost materials considering the above mentioned aspects. In order to remove various types of pollutants the chemical property of activated carbon adsorbent can be changed by varying the ratio of functional groups and carbon atoms fixed in carbonaceous matrix. This preparation method allows the combination of the carbonaceous matrix with other elements in order to obtain composites.

Chapter 5 - Charcoal meal test is widely used for the measurement of gastrointestinal transit in small rodents. Here the authors described some results obtained with the activation of muscarinic, adrenergic and tachykinin NK 2 receptors on gastrointestinal propulsion in conscious rats. Charbachol, reserpine and the selective NK 2 receptor agonist, [βAla8] NKA-(4-10), produced a similar increase of the small intestinal transit in terms of 75-80 % travelled by charcoal as compared to values of 60 % recorded in controls. As expected atropine reverted the prokinetic effect of carbachol. The effect of [βAla8] NKA-(4-10) was restricted to the small intestine since gastric emptying was not affected by [βAla8]NKA-(4-10) and was atropine-resistant. Charcoal test is reliable and has been one of the first experimental models contributing to better understand the role of endogenous tackykinins in the gut motility.

In: Charcoal
Editor: Eric N. Tyrone

ISBN: 978-1-62808-663-8

Chapter 1

CARBON IN ORGANIC PROCESSES: SUPPORT OR CATALYST?

***Yoshiharu Mitoma*[1] *and Cristian Simion*[2]**

[1]Department of Environmental Sciences,
Faculty of Life and Environmental Sciences,
Prefectural University of Hiroshima, , Shobara City, Hiroshima, Japan
[2]Politehnica University of Bucharest, Department of Organic Chemistry,
Bucharest, Romania

ABSTRACT

Since the oldest ages, carbon has been a part of man's history. Since its grinding for making the wonderful drawings of Grotte Chauvet, in the French province of Ardeche, to the breathtaking advances in the field of nanocarbon tubes and graphene, carbon has been in the center of man's scientific preoccupations. Isn't organic chemistry defined as "the chemistry of carbon"?

In the field of organic synthesis, C has been initially used as a support for various metallic catalysts. An extremely useful application in the field of environmental protection was the hydrodechlorination reaction (the replacement of a chlorine atom on a substrate by a hydrogen atom) allowing a dramatic lowering of toxicity. This is a very sensitive subject, since the most toxic compounds to man and to environment, identified by the Stockholm Convention on Persistent Organic Pollutants as the "The Dirty Dozen", are all organochlorinated species, either used or in use as pesticides (DDT, Mirex, Toxaphen, Alachlor, Chlordan,

Aldrin, Dieldrin, Hexachlorcyclohexan, Heptachlor), insulating oil (polychlorobiphenyls) or plainly toxic by-products in synthetic or pyrolitic processes (PCDDs and PCDFs).

We wish to present a review of recent applications of carbon-supported metal catalysts in the field of hydrodechlorination along with a shift that occurred in the past decade in scientists' perception of carbon, since there are more and more reports of various nanoforms of carbon acting solely as catalyst.

1. INTRODUCTION

Over 10,000 years of human collectivities history, waste management problems evolved from a mere nuisance to a dramatic issue. From simple dumping (unfortunately still in use today) to incineration, the road of waste management was and still is a bumpy one [1]. Nowadays, incineration of industrial and municipal wastes is the most common procedure for treating worldwide billions of tons of detrituses. This isn't going without the generation of new and specific problems, such as the *de novo* synthesis of harmful organic volatile pollutants, *e.g.* chlorinated compounds [2] and, among them, those belonging to the dioxin family [3].

Even since the first mention of existing dioxins in the fumes, exhaust gas and in the resulting fly ash of municipal waste treatment plants [4], researchers all around the world have declared war against these extremely toxic compounds, devising a series of detoxification procedures based upon the stripping of chlorine atoms from their structures and replacement with a hydrogen atom in a general process known as *hydrodechlorination* [5]. Thus, numerous reagents, as it or mixtures of reagents, acting as hydrogen donors and in the same time as chlorine acceptors were devised. A simplified scheme of such process could be presented as following:

$$\text{R—Cl} + \text{H—D—Acc} \longrightarrow \text{R—H} + \text{D—Acc—Cl}$$

D = hydrogen donor
Acc = chlorine acceptor (scavenger)

Scheme 1. General outline of a hydrodechlorination process.

In such a process, the chlorinated compound interacts with a hydrogen donor (H-D-Acc) that can act equally as a chlorine acceptor.

Among the various effective procedures, one the most reliable proved to be metal-catalyzed hydrodechlorination (usually a noble metal: Pd, Pt, Rh or Ru) [6], case in which the hydrogen donor usually is molecular hydrogen, but various other hydrogen donors (mostly liquids: solvents, alcohols, even water).

$$R{-}Cl \quad + \quad H_2 \xrightarrow{\text{cat.}} R{-}H \quad + \quad HCl$$

Scheme 2. General outline for catalytic hydrodechlorination process.

Although various metals have been used, a common factor among all these methodologies was the use of carbon as a support.

2. Carbon-Supported Metallic Catalysts and their Applications in Hydrodechlorination Processes

2.1. Palladium [7] and Other Metallic Catalysts on Normal or Activated C Support

Carbon has been known for some time as a preferred choice for metal support in heterogeneous catalytic processes. The preparation of such catalytic systems has been reviewed several times, a comprehensive literature survey being found in Serp and Figueiredo's "*Carbon Materials for Catalysis*" [8]. Even if main procedures involve usually impregnation, adsorption or deposition precipitation, there are newer techniques specifically designed for hydrodechlorination processes. Indeed, as it can be observed from Scheme 2, in these process large amounts of HCl are released, that being the main reason of the advantages of carbon support over silica, alumina, clays and others, since it is better suited to successfully resist these corrosive reaction conditions. However, during some studies of gas-phase hydrodechlorination of dichloromethane with commercial and laboratory-made Pd on activated carbon catalysts, the catalysts suffered severe deactivation after only a few hours of operation [9]. There is not an unanimous conclusion upon deactivation's causes, but most suggested reasons involved either HCl/Cl poisoning, the loss of metal through the formation of chlorides, metal sintering or the formation of carbonaceous deposits [7,10].

Nevertheless, the preparation of the carbon support in itself is an important step of the whole hydrodechlorination procedure. Indeed, the performances of a hydrodechlorination metallic catalyst could be enhanced if commercial active carbon suffers a pretreatment consisting of washing with aqueous NaOH, aqueous HCl and water [11]. Other researchers (involved in the synthesis of ammonia) observed that a serious heat treatment of active carbons, at high temperatures (over 1773K), are also effective for the purification of industrial active carbons [12]. Such pretreatments serves to removes nearly all more or less volatile contaminants, especially typical catalytic poisons like S, As and P. On the other hand, these pretreatments lowers the specific surface area and the carbon support presents a less developed pore structure (compared to thermally untreated carbons), but this shortcoming may be improved by subsequent gasification using steam or carbon dioxide [13].

If such pretreatment proved to be successful for C support used in ammonia synthesis, its application toward hydrodechlorination catalysts, such as using a noble metal as Pd on Norit GF40 carbon was equally effective [14] in the transformation of CCl_2F_2 in CH_2F_2 with up to 90% selectivity. In this study, the C was pretreated, then impregnated and subsequently activated. The study of such process is extremely important since it's a delicate balance between two very actual environmental problems: global warming *versus* ozone layer [15]. Indeed, if CCl_2F_2 is not trapped, treated or destroyed, it is the main gas responsible for ozone layer depletion. If, upon treatment, difluoro-dichloromethane is completely hydrodehalogenated, it's yielding methane, CH_4, one of the major gases that cause global warming.

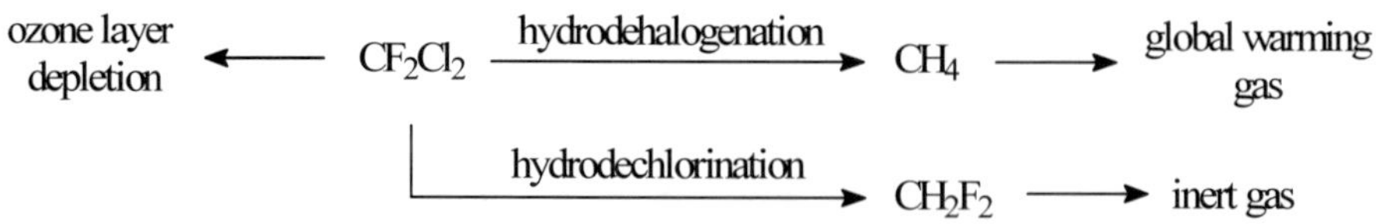

Therefore, numerous studies were dedicated to this subject [10h,11], and, among them there was one in particular who stressed the importance of carbon support [16]. The authors observed that, although the rates and selectivity were similar on all treated samples, Pd black samples had a higher selectivity toward the more hydrogenated products than Pd on activated carbon.

In another study, the Pd/C catalyst used in the gaseous hydrodechlorination of the usual solvent dichloromethane (CH_2Cl_2) was submitted to a much simpler treatment: thermal activation at 250°C, after impregnation of commercial activated carbon [17]. Similar results were obtained for the hydrodechlorination of 1,2-dichloroethane over a dual Pd-Ni catalyst supported on Sibunit carbon [18], tetrachloroethylene, in water/methanol [19] or chlorotrifluoroethylene over usual Pd/C [20].

The common feature of these processes is that they are all hydrodechlorination of small molecules, using a treatment of the substrate in a H_2 gaseous environment. However, most of the organochlorine pollutants are found in solid (soil, fly ash) or liquid (waste water or organic solvent resulted after extraction) matrixes. The hydrodechlorination techniques have diversified to permit the treatment of chlorobenzenes, chlorophenols and other equally toxic compounds in a liquid media, water or an organic solvent, the latter sometimes acting also as a hydrogen donors. Thus, 4-chlorophenol could be successfully treated on Pd/C in a water/formic acid mixture, alcohol or only water [21], while 2-chloro, 4-chlorophenol and 2,4-dichlorophenol were hydrodechlorinated using Pd on mesoporous silica-carbon nano-composite in MeOH/Et_3N or even simply aqueous solutions [22], as well as other chlorinated aromatic chlorides [23]. Even members of more toxic families, such as polychlorinated biphenyls, treated in methanol [24a] or during reductive pyrolysis [24b], octa-chloro-dibenzo-*p*-dioxin (OCDD), in either ethanol-water [25a] or isopropanol-water [25b] mixtures or pesticides, *e.g.* the herbicides Diuron [26] or Alachlor [27], in aqueous solution, were successfully hydrodechlorinated using Pd/C.

Even hybrid processes have been experimented: liquid environment (ethanolic basic solution or aqueous base-*iso*-octane system, at 50°C) and atmospheric pressure of hydrogen, establishing that the reactivity of halides is following the order PhCl > PhBr > PhI in the biphasic system used [28].

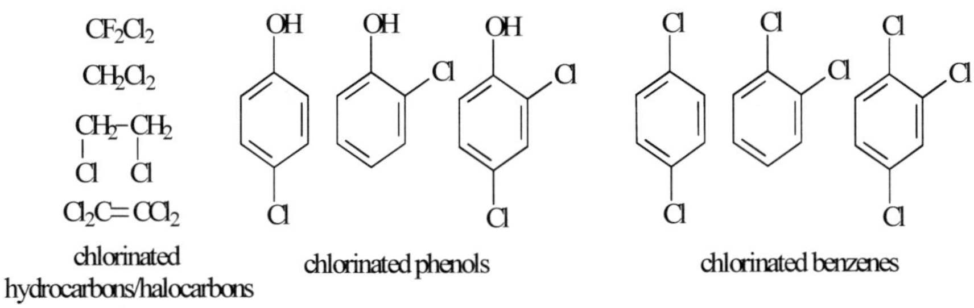

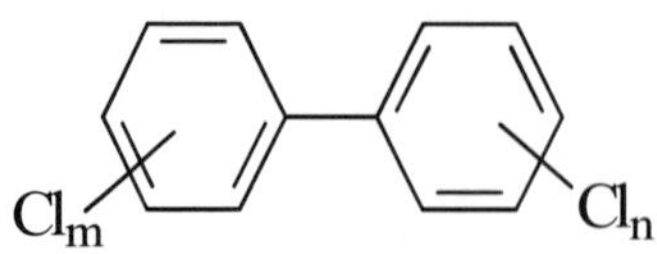

polychlorinated biphenyls

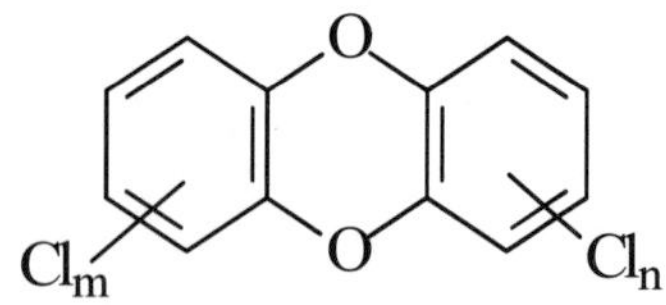

polychloro-dibenzo-*p*-dioxin

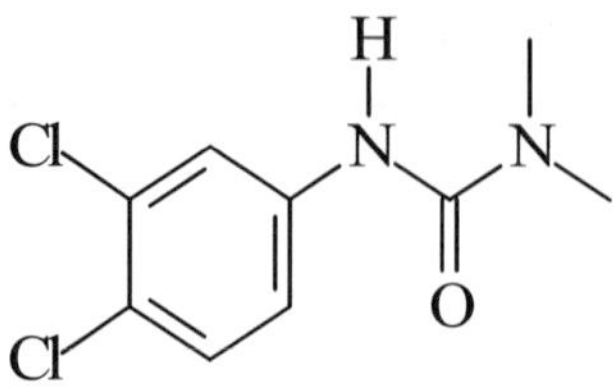

Diuron

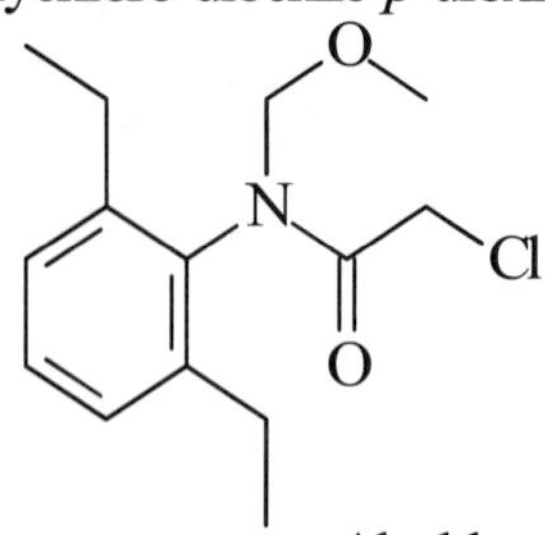

Alachlor

An interesting survey, up to 2005, of hydrodechlorination processes that have occurred on a noble metal Pd/C catalyst can be found in an Amorim and Keane paper [29].

On the other metals that have been used as carbon-supported catalysts in hydrodechlorination processes, mentions can be made for Pt [30], Rh [31], Ru [32], Ni [27, 33], Cu [27] or Fe [34]. A comparison of Pd over Rh reactivity in an aqueous-phase hydrodechlorination of 4-chlorophenol showed that Pd/C offers better yields, while Rh/C is more selective, yielding higher yields of hydrodechlorination/reduction product, cyclohexanol respectively [35]. This result was confirmed by earlier studies of Mitoma and coworkers [36] and was in turn confirmed later by Martin-Martinez *et all.* [37].

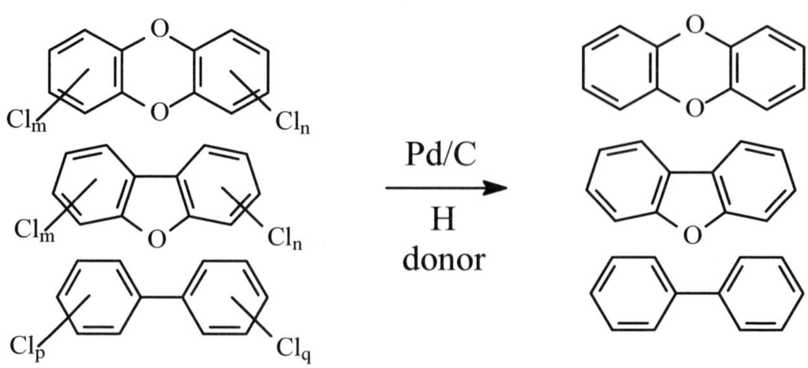

Scheme 3. Hydrodechlorination processes of PCDDs, PCDFs and PCBs.

The Spanish authors evaluated the activities of noble metals such as Pd, Rh, Pt and Ru over activated C and they stated that: "The intrinsic activity or turnover frequency (TOF) of the catalysts follows the order Pd/C > Rh/C > Pt/C > Ru/C in the HDC of dichloromethane (DCM) while the sequence Pd/C > Pt/C > Rh/C > Ru/C was found for the HDC of chloroform (TCM). High selectivities to nonchlorinated products were obtained in all cases except for the HDC of TCM with Rh/C and Ru/C where the selectivity to DCM greatly depends on the operating conditions. A wider diversity of non-chlorinated hydrocarbons was obtained as reaction products with these two catalysts, especially in the HDC of TCM, favoring the formation of carbonaceous deposits which provoked a marked deactivation of the catalysts. In contrast, CH_4 was the only non-chlorinated product obtained with the Pt/C catalysts which showed by far the highest stability." [37].

2.2. Palladium and Other Metallic Catalysts on Structured C Support

All the above-mentioned processes have in common the fact that, with few exceptions, the catalyst is deposed upon commercially available activated carbon. A turning point appears in the 2005 Amorim and Keane's paper [29], where the hydrodechlorination results obtained when working on amorphous and structured carbon support are presented. The conclusion was the same as previously, when instead of simple, commercially available activated carbon, a supplementary *a priori* or *a posteriori* activated carbon was used, which was better results. Indeed, as Amorim and Keane have observed, when graphite or graphite carbon nanofiber were used as support for Palladium, higher hydrodechlorination yields were recorded. If in Amorim and Keane's paper only two forms of carbon were mentioned, the number of nanocarbon is far greater, and continuously growing.

Indeed, since the discovery of Buckminsterfullerene C60, in 1985, by Kroto [38], the number of discovered structures had rapidly increased. Thus, carbon nanotubes and carbon nanohorns (Iijima, 1991) [39], various other fullerenes [40], carbon nanocones [41], nanoscale carbon toroidal structures [42] and helicoidal tubes [43], Schwarzites [44], Haeckelites [45], as well as the various forms of graphene [46] were successively discovered. These carbon structures could be single-walled or multi-walled; they may have zero, positive or even negative Gaussian curvature (*e.g.* Schwarzites).

At about the same period of the 2005 publication of Amorim and Keane's paper, carbon nanotubes, carbon nanofibers and activated carbon have already been used as effective support for metal particles in catalytic conversion of hydrocarbons [47] or for the oxidation of alcohols [48].

A comprehensive review on the use of carbon nanotubes and nanofibers in catalysis in the same period was made by Serp and coworkers [49], showing that non-microporous carbons, such as carbon nanofibres and high surface area graphites were receiving increased attention as catalyst supports.

While Amorim and Keane continued their comparative study of different form of carbon as support [50], similar lines of research were performed by various teams. For instance, Ordonez and coworkers evaluated the performance of carbon nanofibres, high surface area graphites, and activated carbons as supports of Pd-based hydrodechlorination catalysts [51]. The same team, upon the observation that the main deactivation cause is the catalyst fouling caused by formation of coke deposits, blocking the micropores of activated carbons [52], have prepared carbon nanofibres [53] or high-surface area graphites [54] for supporting palladium as catalyst in hydrodechlorination reactions. The conclusions of these studies were that if "carbon nanofibres-based catalysts showed lower initial activity than the activated carbon-supported catalyst (40% vs. 60% initial conversion) but higher stability (after 25 h on stream, activated carbon-based catalyst is completely deactivated whereas carbon nanofibres-based catalysts show activities of about 15%)." [53], than "Obtained results show that high-surface area graphites-supported catalysts present very good performance both in terms of conversion (TOFs up to four times higher than the encountered for other conventional supports, such as activated carbon or alumina), selectivity (only small amounts of trichloroethylene being observed, in the same order of magnitude, below 10% selectivity in all the cases, that other Pd catalysts), and resistance to deactivation (more than 30 h on stream)." [54]. Therefore, there are obvious differences between the activity of the various forms of nanocarbon. Thus, several other forms were tested, such as various metal-carbon nanocomposites [55], palladium on a carbon-silica nanocomposite [56], carbon nanotubes [57] and even ultradispersed diamond (UDD) [58].

According to the Russian authors, a very high catalytic activity of Pd or even of an unusual hydrodechlorination catalyst, gold Au, supported on ultradispersed diamond (UDD) was observed: "Specifically, Pd/UDD shows remarkable selectivity in the hydrodechlorination of 1,3,5-trichlorobenzene (benzene was the main product already at 30% TCB conversion and became the only product after 30 minutes of the reaction), substantially exceeding the performance of Pd supported on activated carbon and commercial 5% Pd/C references. It is proposed that this behavior is linked to the oxygen containing functional groups at the ultradispersed diamond surface, as detected via IR and EDX/SEM mapping, combined with a very small particle size of ultradispersed diamond support." [58].

The Polish authors, however, propose two reasons for the higher selectivity to the more deeply dehalogenated products: the fine dispersion of the metal inside the support and the trapping of metal particles inside the channels of the support [57]. But this seemed to be true for Pd. The same authors observed that for Ni/C, the behavior of the catalysts seems to be regulated by the population of surface carbon species. Thus, a higher surface carbon content creates "an undeniable difficulty" in removal of the supplementary chlorine atom from 1,2-dichloroethane [59].

Another approach was to treat carbon nanofibers with ammonia, at elevated temperatures, in order to generate pyridinic-type basic species on the carbon nanofibers surface, which will serve as anchors for Pd nanoparticles, resulting in both good dispersion and stability of the catalytic system. Moreover, the basic species will also trap evolved HCl [60].

These specific forms of metal on structured carbon catalysts have proven to be also effective for other hydrogenation process, such as Ni on carbon nanofibers in the selective hydrogenation of 2-*tert*-butylphenol [61] or Pd on detonation nanodiamond in the hydrogenation of multiple bonds (>C=C< and -C≡C-) [62].

An interesting paper was published in 2007 concerning a solvent-free, mechanochemical treatment of solid hexachlorobenzene with commercial Pd/C [63], in the presence of various hydrogen donors. According to the Canadian authors, the presence of a base was needed (NaOH), while mmoles amounts of water was added (introduced in the form of sodium phosphate heptahydrate) as a necessity to improve hydrodechlorination efficiency; other hydrogen donors were also tested, such as sodium hypophosphite, sodium borohydride and calcium hydride. Although the authors have stressed the importance of the proper hydrogen donor, the possible transformation of the carbon [63] support was overlooked.

The overall conclusion that seemed to stand, concerning the efficiency of the various catalytic carbon-supported systems, was related to the shape and area of the carbon substrate [64]. This was confirmed by the attempts of obtaining other type of carbon supports, such as the carbon xerogels [65].

3. Carbon As Adsorbent in Pollution Control

Before being recognized as a valuable support for various metallic catalysts, carbon has been known for his adsorption properties. In relation to the subject of organochlorine pollutants and hydrodechlorination, the intervention of carbon was identified on two levels: retention, respectively dispersion of volatile organic pollutants belonging to the class of dioxins. Indeed, carbon (or charcoal) has found numerous applications as adsorbent for the retention of dioxins, but was also regarded, as biochar, as a carrier in the environment of these same compounds. Activated carbon treatment has been employed for the removal of dioxins because it is a simple and low-cost removal technology.

3.1. Carbon As Adsorbent in Organohalogen Depolluting Technologies

Major sources of dioxin-like congener (PCDD/Fs and PCBs) are (as order of emission's importance): mining and metal processing, waste burning, thermal power plants, industrial boilers, motor transportation and crematoria [66]. All incinerators, either large-scale municipal waste incinerator (MWI), or industrial waste incinerator (IWI) must be provided with air pollution control devices (APCDs), which can be, for example, activated carbon injection with bag filter or fixed activated carbon bed. These activated carbon adsorption devices are used to control PCDD/Fs and PCBs emissions, though partitioning between vapor and solid phases and their subsequent removal efficiencies. Such a study performed for both MWI and IWI showed that average PCDD/Fs and PCBs concentrations of stack gas were 0.031 and 0.006 ng-TEQ/N m^3, when activated carbon injection with bag filter were used, and respectively 1.74 and 0.19 ng-TEQ/N m^3 for fixed activated carbon bed [67].

To facilitate this toxicological risk assessment of PCDDs, PCDFs, and co-PCBs, experts of the World Health Organization developed, starting from the *toxic equivalent quotient* (WHO-TEQ), the *toxic equivalency factors* (WHO-TEF) for these substances. The TEQ system was developed to compare risks of exposure in areas of contamination with widely various compositions and levels of most toxic dioxins, furans, and biphenyls. The toxic action of a substance is weighed against the toxicity of the most toxic dioxin, 2,3,7,8-TCDD (whose TEF is arbitrarily equal to 1.0). Consequently, to each of the 29 highly toxic dibenzodioxins, dibenzofurans, and biphenyls was assigned a specific TEF; partial TEQs are obtained as the product between the individual concentration and the corresponding TEF, and the sum of partial TEQs represent the samples' total TEQ [68].

Thus, these results showed that fixed activated carbon beds could reduce vapor-phase PCDD/Fs and PCBs concentrations in flue gas, while the activated carbon injection with bag filters can effectively adsorb the vapor-phase dioxin-like congener and collect the solid-phase PCDD/Fs and PCBs. From the same researcher came the measurements showing that that each gram activated carbon adsorbs 105–115 ng-PCDD/Fs and each surface area (m^2) of activated carbon adsorbs 10–25 ng-PCDD/Fs, confirming that the surface area of mesopore and macropore (20–200Å) of the activated carbon is a critical factor affecting PCDD/F adsorption capacity [67].

Russian scientists managed to simultaneously lower costs and improve overall retention by mixing carbon in the flue gas (thus, the contact time between carbon and dioxins is considerably longer) [69]. A Chinese team was even able to determine the minimum feeding rate of activated carbon to control dioxin emissions [70]. They have found that dioxin removal efficiency was enhanced by an increase in activated carbon feeding concentration with an almost linear function to activated carbon feeding rate (mg/min) versus volumetric flow rate of flue gas (Nm^3/min). Optimum values were around 65 mg/Nm^3, while no influence was sensed over 150 mg/Nm^3. Removal efficiencies can be raised to almost 100% when the flow gas is passed through a multi-layer reactor [70]. Thus, excellent PCDD/F removal efficiency (>99.99%) was achieved with a multi-layer design with bead shaped activated carbons.

Analyses were carried out in order to measure the major physicochemical properties of activated carbon (micropore volume, pore diameter, mesopore volume, Brunauer-Emmett-Teller (BET) surface area, pH, metal elements or surface acid functional groups) [71]. Most important characteristics for activated carbon to function optimally as adsorbent were a surface area of 700–1200 m^2/g and micropores with diameters of about 0.7–0.8 nm. The mesopore volume of activated carbon also influences the adsorption rate and the equilibrium adsorption [72].

A comparison in the adsorption capacity for volatile organic compounds onto carbon nanotubes, carbon nanofibers, and high-surface-area graphites was made [73], and showed that among these species, high-surface-area graphites present the highest adsorption capacity, followed by carbon nanotubes and carbon nanofibers (although carbon nanotubess presented an intermediate surface area between the two types of high-surface-area graphites studied).

Moreover, carbon nanotubes proved to be the material with the highest dispersive contribution.

Williams and coworkers proved that the source of carbon is unimportant in that matter [74]. These researchers compared the removal efficiency of three types of activated carbons derived from the pyrolysis of refuse derived fuel, textile waste and scrap tire with that of a commercial carbon. Thus, the total PCDD/Fs toxic equivalent removal efficiencies in the exhaust gas stream were 58%, 57%, 64% and 52%, respectively. Analysis of the reacted activated carbons showed that there was some formation of PCDD/F, for instance, a total of 60.6 μg I-TEQ/kg toxic PCDD/F was formed in the refuse derived fuel activated carbon compared to 34 μg I-TEQ/kg in the commercial activated carbon [74]. Therefore, even activated carbon obtained from waste incineration can be useful. These results were later confirmed by a Sino-Japanese team who managed to improve the retention yields by recycling charcoal, enabling it to be use as both a thermal source and an adsorbent [75].

However, there are some concerns with the injection of activated carbon in the gas flow, such as powdered activated carbon could ignite at temperatures above 200°C, and lead to fires and plant downtime [76].

If activated carbon retains easily volatile organohalogen compounds, while metals deposed on active carbon are effective in the hydrodehalogenation of the same volatile organohalogen compounds, the next logical step was to combine these two procedures. Indeed, adversaries of the powdered active carbon injection in gas flow advocated that powdered active carbon adsorbs dioxins, but does not destroy them, leading to the disposal of larger amounts of solid residues (highly contaminated with PCDD/Fs), that requires new operations, new maintenance and even new equipment and technologies and (again) that fire hazard is always present [76].

However, this step of combining adsorption properties of carbon with the catalytic hydrodechlorinating properties of metals on carbon have been already taken, by simply preparing metal-containing adsorbents in which the metal was either noble metal cataysts such as Pd [77], or simple metals such as Fe-Pd [78] or Cu/C and Fe/C [79]. An interesting observation was made by Choi and coworkers, who noticed that the quantity of Pd doesn't need to be excessive – 0.01% (in weight), which was enough to obtain the same hydrodechlorination rates that than in the usual process in which commercially available 5% (in weight) Pd/C was used [78].

3.2. Carbon as a Transport Agent in the Dispersion of Dioxins in Soil

Thus far, carbon was considered as a support for metal or noble metal catalysts, working in the same time as an adsorbent. Best carbon form known to be an adsorbent, with wide applications, is biochar. Biochar is a solid, carbon-rich material obtained from the carbonization of biomass in a closed container with little or no available air. It can be added to soils with the intention to improve agriculture and the environment in several ways, and its stability in soil and superior nutrient-retention properties make it an ideal soil amendment to increase crop yields. In addition, biochar may reduce emissions from biomass that would otherwise naturally degrade to greenhouse gases. Biochar is found in soils around the world as a result of vegetation fires and historic soil management practices. Intensive study of biochar-rich dark earths in the Amazon (*terra preta*), has led to a wider appreciation of biochar's unique properties as a soil enhancer [80].

There is however a certain concern that dioxins, as well as polyaromatic hydrocarbons, may be formed during biochar preparation. These contaminants can possibly be present in the biochar matrix and even bioavailable to exposed organisms. The total and bioavailable polyaromatic hydrocarbons and dioxins were quantified in a suite of over 50 biochars produced via slow pyrolysis between 250 and 900°C, using various methods and biomass from tropical, boreal, and temperate areas [81]. Fortunately, the total dioxin concentrations were low (up to 92 pg/g) and bioavailable concentrations were below the analytical limit of detection.

But due to their extremely good properties as adsorbent, charcoal, activated carbon and even biochar were used to contribute to the decontamination process of polluted soils, by reducing the availability of aged polychlorinated dibenzo-*p*-dioxin/dibenzofurans [82]. On the carbon samples tested in this study, best retention results were obtained using the virgin activated carbon used in food and beverage industry (AquaPAC-S, <45 μm, specific surface area of 1150 m^2/g). Mixed results were obtained using a corn stover-based biochar (CS Biochar, <250 μm, specific surface area of 67.2 m^2/g) while worst results were recorded when coconut-based AC (Coconut AC, 75−300 μm, specific surface area of 1320 m^2/g) was used [82].

On the other hand, if charcoal may adsorb dioxins from soil, it does not mean that it will immobilize it. There were studies dealing with the role that such entities will play in the distribution in soil of organohalogen compounds [83]. Thus, in the attempt to explain the elevated levels of polychlorinated dibenzo-*p*-dioxins/dibenzofurans in some floodplain soils, the authors first identified distinct congener patterns consistent with byproducts from either chloralkali manufacturing or chlorophenols productions. The question was: how were these compounds transported there? The answer became obvious when black carbon particles were isolated for the first time from floodplain soil impacted by PCDD/Fs. Petrographic analyses showed that black carbon particles, including coal, oxidized coal, metallurgical coke, depositional carbon, coal tar/pitch, cenosphere, and charcoal, comprised approximately 30% by volume of the organic fraction with size range of 250 μm-2000 μm from a typical floodplain soil. The black carbon particles with anthropogenic origin such as pitch and coke associated with the chloralkali production process served as both the source and subsequent transporter for the highly hydrophobic PCDD/Fs. These anthropogenic black carbon particles were enriched with high levels of PCDFs, containing approximately 1000-fold the concentration found in the bulk soil. The strong association of PCDD/Fs with anthropogenic black carbon directly impacts the physicochemical and biological availability, thus the risk associated with these hydrophobic organochlorines in soils and sediments.

4. Carbon As Single Catalyst

At this point the question "*what is the exact role that carbon is playing during hydrodechlorination process?*" seems a legitimate one. It is sure that carbon is a most valuable adsorbent for all types of chlorinated compounds. It also well-documented that carbon is a worthy support for several types of metallic catalysts used in hydrodechlorination process, more and more reports underlying the importance of structured carbon, either as carbon nanotubes, carbon nanofibers, graphite or even diamond. However, the observation made by Choi and coworkers, who noticed that a mere 0.01% (in weight) Pd is enough to obtain the same hydrodechlorination rates that for commercially available 5% (in weight) Pd/C [78] was an intriguing one.

When analyzing their characteristics, catalysts embedded in a carbon matrix offer better efficiency even at moderate temperatures due to their extremely porous structure, affording thus a large specific surface area. Of utmost importance for the intervention of carbon as a catalyst is the aspect that due to the high density of reactive edge planes in these graphitic carbon materials, the activated reactant species that are formed on their surfaces are immobilized enough time to react with the substrates [84].

It has been known for some time that, besides being an efficient catalytic support, carbon supports can function as active catalyst themselves. This was demonstrated in several cases of oxidative dehydrogenation, such as the transformation of ethylbenzene to styrene [85]. In this paper, the authors compared the performances (in the oxidative dehydrogenation of ethylbenzene) of multi-walled carbon nanotubes to those of activated carbon and graphite samples.

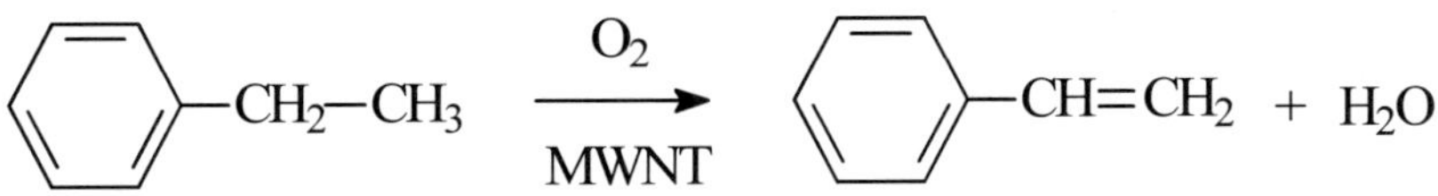

Multi-walled carbon nanotubes were not the only carbon forms used to catalyze this process. Highly graphitized multi-walled carbon nanofibers were also successfully tested [86], as well as nanodiamonds [87]. Other equally successful oxidative dehydrogenation processes were applied to lower alkanes, such as butane [88], when the use of single-walled or multi-walled carbon nanotubes yielded a mixture of butenes and butadiene, but also large amounts of CO and CO_2), again oxidative dehydrogenation of butane on glassy carbon spheres, single-walled carbon nanotubes, graphene, and graphite [89], of propane, on carbon nanofibers [90], 9,10-dihydroanthracene using multi-walled carbon nanotubes [91] or that of carbon nanoparticles immobilized on zeolites to the transformation of vinylcyclohexene to styrene [92]. An interesting aspect in the Korean team's paper [89] was the fact that they also used Pd-containing carbon nanofibers. They noticed that noble metals such as Pd-containing carbon nanofibers presented a superior selectivity for butadiene formation than Pd-free carbon nanofibers catalyst. More interesting was that carbon with graphitic structures was highly selective for propylene and butenes without severe combustion in the oxidative dehydrogenation of butane.

However, in another manuscript, Russian scientists who investigated the catalytic conversion of C2–C3 alcohols on detonation nanodiamond observed that the oxidizing and reducing treatment of nanodiamond changes its activity and selectivity, and the activity of oxidized nanodiamond is considerably higher than that of reduced nanodiamond [93]. For its part, graphene (as graphene oxide) is a very suitable catalyst in the mild and efficient oxygenation of benzyl alcohol to benzaldehyde, with a conversion of over 90% [94].

It is of common acceptance that the success of nanocarbon catalysts in oxidative dehydrogenation processes is due to the possibility of both sp^2 forms (carbon nanotubes, carbon nanofibers or graphene) and sp^3 forms (nanodiamond) to link oxygen [95], either in quinoid moieties [96], or as epoxy or even hydroxyl functional groups [97]. There is than no surprise if oxidative purification of carbon nanotubes is enhancing its catalytic performance in oxidative dehydrogenation reactions [98]. According to the authors, oxidative purification with mild diluted HNO_3, followed by NaOH washing, lowers the amount of amorphous carbon attached to multi-walled carbon nanotubes. This kind of "activation" is paving the way to oxygen fixation by the more available regulate surfaces of the multi-walled carbon nanotubes.

5. Carbon As Single Catalyst in Hydro-dechlorination Processes

Thus, an entire new research field is created, terms are coined (such "carbocatalysis") and all eyes are expectedly and hopefully staring into the future, the so-called imminent "Carbon age" [99]. Meanwhile the applications of various nanocarbon forms are widening: a comprehensive review of these new applications was recently published by Chenliang Su and Kian Ping Loh [100].

But what all these new applications have in common is that they are situated on the oxidative side of organic synthesis and organic processes. What about its applications in the register of hydrodechlorination reactions? As surprising as it seems, there are reports of such processes.

A few years back, Kakuta and coworkers studied the *de novo* synthesis of PCDD/Fs in fly ash from municipal solid waste incinerators [101]. These researcher concluded that unburned carbon in solid waste and activated carbon powder injected into flue gas were the main carbon sources in fly ash and that the PCDD/F formation by *de novo* synthesis depended on gasifiable carbon at the temperature of 450°C, not on the total residual carbons, regardless of the origin of the carbon. Moreover, the yield of PCDD/Fs by *de novo* synthesis was also influenced by metal content in fly ash (more specifically Cu,in that case).

These papers were abundantly provided with convincing proofs, both theoretical and experimental. But how is the formation of PCDD/Fs is possible, in the light of all precedent discussions concerning either their adsorption on C, either their destruction when metals are present? Was it only those hydrogen donors are obviously absent in a furnace? On the other hand, in the early 2000s, a patent was submitted by researchers from the United States Department of Energy concerning the formation of olefins from chlorinated substrates, through contact with diamonds for a sufficient time and at a sufficient temperature, *via* elimination reactions [102].

Ghaffar and Tabata attempted to answer the previous question using fly ash as a catalytic substrate to achieve hydro-dechlorination, in the presence of hydrogen donors [103]. After understanding the adverse role played by the metal content in fly ash and by the need of milder conditions [103], Ghaffar and Tabata succeeded in the first hydrodechlorination of an aromatic chloroderivative (*e.g.* 2,4,5-trichlorobiphenyl) in the presence of only activated carbon and of a hydrogen donor (a reducing medium of calcium hydroxide/sodium hydroxide and sulfur in water/isopropanol solution) [104].

CONCLUSION

As a conclusion, it is more than obvious that carbon has stepped in new era, where most of the nanocarbons different forms have demonstrated to exhibit a comparable, or even better, catalytic performance and long-term stability for various organic reactions [105]. Along with its use as a support for pure and heterogeneous metal catalysis, carbon has also been demonstrated to show interesting catalytic activities, playing this role disguised in a variety of nanostructured carbons, such as carbon nanotubes, carbon nanofibers, carbon onions, nanodiamonds and graphene, which all have been found to efficiently catalyze various organic processes, including hydrodechlorination ones, and are all belonging to a present rich class of solid-state materials that are nonpolluting and re-usable. Thus, Kekulé's old definition stating that "*organic chemistry is the chemistry of carbon*" (*Lehrbuch der Organischen Chemie*, 1861, Vol. 1, p. 11) has just started to gain a completely new perspective.

REFERENCES

[1] M.F. Leman in "*Waste Management*", Peter Lang A.G., International Academic Publishers, Bern, 2008.

[2] a) S. Ramamoorthy, S. Ramamoorthy in "*Chlorinated Organic Compounds in the Environment*", CRC-Press, Boca Raton, Fl., 1997; b) R. Stringer, P. Johnston in "*Chlorine in the Environment: An Overview of the Chlorine Industry*", Kluwer Academic Publishers, Dordrecht, 2001; c) J. Thornton in "*Pandora's Poison: Chlorine, Health and a New Environmental Strategy*", MIT Press, London, 2000.

[3] L.M. Gibbs in "*Dying from Dioxin: A Citizen's Guide to Reclaiming Our Health and Rebuilding Democracy*", South End Press, Cambridge, 1995; b) T. Pollock in "*Dioxins and Furans: Questions and Answers*", Academy of Natural Sciences, Philadelphia, 1989; c) E.W. Bretthauer, H.W. Kraus in "*Dioxin Perspectives: A Pilot Study on International Information Exchange on Dioxins and Related Compounds (NATO Challenges of Modern Society)*", Plenum Publications Corp., New York, 1992; d) R. Allen in "*The Dioxin War: Truth and Lies about a Perfect Poison*", Pluto Press, London, 2004; e) A. Schecter, T.A. Gasiewics in "*Dioxin and Health*", Wiley-Interscience, New York, 2003; f) D.R. Zook, C. Rappe in "*Dioxins and Health-Environmental Sources, Distributionand Fate of Polychlorinated Dibenzodioxins, Dibenzofurans and Related Organochlorines*", Schecter, A., Plenum, New York, 1994.

[4] K. Olie, P.L. Vermeulen, O. Hutzinger, Chlorodibenzo-*p*-dioxins and chlorodibenzofurans are trace components of fly ash and flue gas of some municipal waste incinerators in the Netherlands, *Chemosphere*, 1977, *6(8)*, 455-459; b) J.W.A. Lustenhouwer, K. Olie, O. Hutzinger, Chlorinated dibenzo-*p*-dioxins and related compounds in incinerator effluents, *Chemosphere*, 1980, *9*, 501-522; c) O. Hutzinger, M.J. Blumich, Sources and fate of PCDDs and PCDFs: an overview, *Chemosphere*, 1985, *14*, 581-600.

[5] B.F. Hagh, D.T. Allen in "*Innovative hazardous waste treatment technology*", vol. 1, H.M. Freedman ed., Technomics, Lancaster, Pa., 1990.

[6] F. Alonso, I.P. Beletskaya, M. Yus, Metal-mediated reductive hydrodehalogenation of organic halides. *Chem Rev*, 2002, *102*, 4009-4091, and references cited therein.

[7] F.J. Urbano, J.M. Marinas, Hydrogenolysis of organohalogen compounds over palladium supported catalysts, *J Mol Catal A Chem*, 2001, 173, 329-345.

[8] J.H. Bitter, K.P. de Jong, Preparation of Carbon-Supported Metal Catalysts, in "*Carbon Materials for Catalysis*", P. Serp and J. L. Figueiredo eds., John Wiley & Sons, Inc., Hoboken, NJ, 2008.

[9] a) Z.M. de Pedro, L.M. Gomez-Sainero, E. Gonzalez-Serrano, J.J. Rodriguez, Gasphase hydrodechlorination of dichloromethane at low concentrations with palladium/carbon catalysts, *Ind Eng Chem Res*, 2006, 45, 7760-7766; b) M.A. Álvarez-Montero, L.M. Gómez-Sainero, M. Martín-Martínez, F. Heras, J.J. Rodriguez, Hydrodechlorination of chloromethanes with Pd on activated carbon catalysts for the treatment of residual gas streams, *Appl Catal B Environ*, 2010, 96, 148-156.

[10] a) C.A. Gonzalez, M. Bartoszek, A. Martin, C. Montes de Correa, Hydrodechlorination of light organochlorinated compounds and their mixtures over Pd/TiO_2-washcoated minimonoliths, *Ind Eng Chem Res*, 2009, *48*, 2826-2835; b) P. Forni, L. Prati, M. Rossi, Catalytic dehydrohalogenation of polychlorinated biphenyls Part II: Studies on a continuous process, *Appl Catal B Environ*, 1997, *14*, 49-53; c) S. Ordonez, F.V. Diez, H. Sastre, Characterisation of the deactivation of platinum and palladium supported on activated carbon used as hydrodechlorination catalysts, *Appl Catal B Environ*, 2001, *31*, 113-122; d) B. Heinrichs, F. Noville, J. Schoebrechts, J. Pirard, Palladium-silver sol-gel catalysts for selective hydrodechlorination of 1,2-dichloroethane into ethylene: IV. Deactivation mechanism and regeneration, *J Catal*, 2003, *220*, 215-225; e) D. Chakraborty, P.P. Kulkarni, V.I. Kovalchuk, J.L. d'Itri, Dehalogenative oligomerization of dichlorodifluoromethane catalyzed by activated carbonsupported Pt-Cu catalysts: effect of Cu to Pt atomic ratio, *Catal Today*, 2004, *88*, 169-181; f) M. Legawiec-Jarzyna, A. Srebowata, W. Juszczyk, Z. Karpinski, Hydrodechlorination of dichlorodifluoromethane, carbon tetrachloride and 1,2-dichloroethane over Pt/Al2O3 catalysts, *J Mol Catal A Chem*, 2004, 224, 171-177; g) N.C. Concibido, T. Okuda, W. Nishijima, M. Okada, Deactivation and reactivation of Pd/C catalyst used in repeated batch hydrodechlorination of PCE, *Appl Catal B Environ*, 2007, *71*, 64-69; h) M. Makkee, E.J.A.X. van de Sandt, A. Wiersma, J.A. Moulijn, Development of a satisfactory palladium on activated carbon catalyst for the selective hydrogenolysis of CCl_2F_2 (CFC-12) into CH_2F_2 (HFC-32), *J Mol Catal A Chem*, 1998, *134*, 191-200; i) D. Ju Moon, M. Jo Chung, K. You Park, S. In Hong, Deactivation of Pd catalysts in the hydrodechlorination of chloropentafluoroethane, *Appl Catal A Gen*, 1998, *168*, 159-170; j) E.S. Lokteva, T.N. Rostovshchikova, S.A. Kachevskii, E.V. Golubina, V.V. Smirnov, A.Y. Stakheev, N.S. Telegina, S.A. Gurevich, V.M. Kozhevin, D.A. Yavsin, High catalytic activity and stability of palladium

nanoparticles prepared by the laser electrodispersion method in chlorobenzene hydrodechlorination, *Kinet Catal*, 2008, *49*, 748-755.

[11] E.J.A.X. van de Sandt, A. Wiersma, M. Makkee, H. van Bekkum, J.A. Moulijn, Selection of activated carbon for the selective hydrogenolysis of CCl_2F_2 (CFC-12) into CH_2F_2 (HFC-32) over palladium-supported catalysts, *Appl Catal A*, 1998, *173*, 161-173.

[12] a) L. Forni, D. Molinari, I. Rossetti, N. Pernicone, Carbon-supported promoted Ru catalyst for ammonia synthesis, *Appl Catal A*, 1999, *185*, 269-275; b) X. Zheng, S. Zhang, J. Xu, K. Wei, Effect of thermal and oxidative treatments of activated carbon on its surface structure and suitability as a support for bariumpromoted ruthenium in ammonia synthesis catalysts, *Carbon*, 2002, *40*, 2597-2603.

[13] Z. Kowalczyk, S. Jodzis, W. Rarog, J. Zielin□ski, J. Pielaszek, A. Presz, Carbon-supported ruthenium catalyst for the synthesis of ammonia. The effect of the carbon support and barium promoter on the performance, *Appl Catal A*, 1999, *184*, 95-102; b) Z. Kowalczyk, J. Sentek, S. Jodzis, R. Diduszko, A. Presz, A. Tetrzyk, Z. Kucharski, J. Suwalski, Thermally modified active carbon as a support for catalysts for NH_3 synthesis, *Carbon*, 1996, *34(3)*, 403-409.

[14] M. Bonarowska, W. Rarog-Pilecka, Z. Karpinski, The use of active carbon pretreated at 2173K as a support for palladium catalysts for hydrodechlorination reactions, *Catal Today*, 2011, 169, 223-231.

[15] B.O. Bolaji, Z. Huan, Ozone depletion and global warming: Case for the use of natural refrigerant – a review, *Renew Sustain Energ Rev*, 2013, *18*, 49-54.

[16] R.M. Rioux, C.D. Thompson, N. Chen, R.H. Ribeiro, Hydrodechlorination of chlorofluorocarbons CF_3-$CFCl_2$ and CF_3-CCl_3 over Pd/carbon and Pd black catalysts, *Catal Today*, 2000, *62*, 269-278.

[17] a) L.M. Gomez-Sainero, X.L. Seoane, J.L.G. Fierro, A. Arcoya, Liquid-phase hydrodechlorination of CCl_4 to $CHCl_3$ on Pd/carbon catalysts: nature and role of Pd active species, *J Catal*, 2002, *209*, 279-288; b) M.A. Alvarez-Montero, L.M. Gomez-Sainero, J. Juan-Juan, A. Linares-Solano, J.J. Rodriguez, Gas-phase hydrodechlorination of dichloromethane with activated carbon-supported metallic catalysts, *Chem Eng J*, 2010, 162, 599-608; c) Z.M. de Pedro, J.A. Casas, L.M. Gomez-Sainero, J.J. Rodriguez, Hydrodechlorination of dichloromethane with a Pd/AC catalyst: Reaction pathway and kinetics, *Appl Catal B Environ*, 2010, *98*, 79-85; d) M.A. Alvarez-Montero, L.M. Gomez-Sainero, M. Martın-Martinez, F. Heras, J.J. Rodriguez,

Hydrodechlorination of chloromethanes with Pd on activated carbon catalysts for the treatment of residual gas streams, *Appl Catal B Environ*, 2010, *96*, 148-156.

[18] A. Srebowata, W. Juszczyk, Z. Kaszkur, Z. Karpinski, Hydrodechlorination of 1,2-dichloroethane on active carbon supported palladium–nickel catalysts, *Catal Today*, 2007, 124, 28-35.

[19] N.C. Concibido, T. Okuda, Y. Nakano, W. Nishijima, M. Okada, Enhancement of the catalytic hydrodechlorination of tetrachloroethylene in methanol at mild conditions by water addition, *Tetrahedron Lett*, 2005, *46*, 3613-3617.

[20] B.-C. Meng, Z.-Y. Sun, J.-P. Ma, G.-P. Cao, W.-K. Yuan, Selective liquid-phase hydrodechlorination of chlorotrifluoroethylene over palladium-supported catalysts: activity and deactivation, *Catal Lett*, 2010, *138*, 68-75.

[21] a) L. Calvo, M.A. Gilarranz, J.A. Casas, A.F. Mohedano, J.J. Rodrıguez, Hydrodechlorination of 4-chlorophenol in water with formic acid using a Pd/activated carbon catalyst, *J Haz Mat*, 2009, 161, 842-847; b) C. Xia, Y. Liu, S. Zhou, C. Yang, S. Liu, S. Guo, Q. Liu, J. Yu, J. Chen, The influence of ion effects on the Pd-catalyzed hydrodechlorination of 4-chlorophenol in aqueous solutions, *Catal Commun*, 2009, *10*, 1443-1445; c) C. Xia, Y. Liu, J. Xu, J. Yu, W. Qin, X. Liang, Catalytic hydrodechlorination reactivity of monochlorophenols in aqueous solutions over palladium/carbon catalyst, *Catal Commun*, 2009, *10*, 456-458; d) Y. Ukisu, Highly enhanced hydrogen-transfer hydrodechlorination and hydrogenation reactions in alkaline 2-propanol/methanol over supported palladium catalysts, *Appl Catal A Gen*, 2008, 349(1-2), 229-232.

[22] Z. Jin, C. Yu, X. Wang, Y. Wan, D. Li, G. Lu, Liquid phase hydrodechlorination of chlorophenols at lower temperature on a novel Pd catalyst, *J Haz Mat*, 2011, *186*, 1726-1732; b) Y. Shao, Z. Xu, H. Wan, Y. Wan, H. Chen, S. Zheng, D. Zhu, Enhanced liquid phase catalytic hydrodechlorination of 2,4-dichlorophenol over mesoporous carbon supported Pd catalysts, *Catal Commun*, 2011, *12*, 1405-1409.

[23] Y. Monguchi, A. Kume, K. Hattori, T. Maegawa, H. Sajiki, Pd/C–Et3N-mediated catalytic hydrodechlorination of aromatic chlorides under mild conditions, *Tetrahedron*, 2006, *62*, 7926-7933; b) Tadeusz Janiak, Janina Okal, Effectiveness and stability of commercial Pd/C catalysts in the hydrodechlorination of meta-substituted chlorobenzenes, *Appl Catal B Environ*, 2009, *92*, 384-392.

[24] S. Ishihara, A. Ido, Y. Monguchi, H. Nagase, H. Sajiki, Pd/C-catalyzed dechlorination of polychlorinated biphenyls under hydrogen gas-free conditions, *J Haz Mat*, 2012, *229-230*, 15-19; b) E. Lorenc-Grabowska, J. Yperman, G. Gryglewicz, S. Hoste, R. Carleer, Study on hydrodechlorination of polychlorinated biphenyls during reductive pyrolysis in the presence of a catalyst, *Fuel*, 2006, *85*, 374-381.

[25] F. Zhang, J. Chen, H. Zhang, Y. Ni, X. Liang, The study on the dechlorination of OCDD with Pd/C catalyst in ethanol–water solution under mild conditions, *Chemosphere*, 2007, *68*, 1716-1722; b) Z. Yang, C. Xia, Q. Zhang, J. Chen, W. Wu, X. Liang, A. Kettrup, Treatment of PCDD/Fs and PCBs in fly ash extracts under mild conditions, *Fresenius Environ Bull*, 2006, *15(2)*, 86-94.

[26] L. Calvo, M.A. Gilarranz, J.A. Casas, A.F. Mohedano, J.J. Rodriguez, Hydrodechlorination of diuron in aqueous solution with Pd, Cu and Ni on activated carbon catalysts, *Chem Eng J*, 2010, *163*, 212-218.

[27] L. Calvo, M.A. Gilarranz, J.A. Casas, A.F. Mohedano, J.J. Rodriguez, Hydrodechlorination of alachlor in water using Pd, Ni and Cu catalysts supported on activated carbon, *Appl Catal B Environ*, 2008, *78*, 259-266.

[28] S.S. Zinovyev, A. Perosa, P. Tundo, Liquid-phase and multiphase hydrodehalogenation of halobenzenes over Pd/C: Reaction selectivity and inhibition/promotion effects by the quaternary salt, *J Catal*, 2004, *226*, 9-15.

[29] C. Amorim, G. Yuan, P.M. Patterson, M.A. Keane, Catalytic hydrodechlorination over Pd supported on amorphous and structured carbon, *J Catal*, 2005, *234*, 268-281, and references cited therein.

[30] a) M.A. Alvarez-Montero, L.M. Gomez-Sainero, A. Mayoral, I. Diaz, R.T. Baker, J.J. Rodriguez, Hydrodechlorination of chloromethanes with a highly stable Pt on activated carbon catalyst, *J Catal*, 2011, *279(2)*, 389-396; b) A.J.L. Benítez, G. Del Angel, Total hydrodechlorination of industrial transformer oil on metal-supported catalysts, *Chem Eng Commun*, 2009, *196(10)*, 1217-1226; c) T. Yoneda, T. Takido, K. Konuma, Hydrodechlorination reactivity of para-substituted chlorobenzenes over platinum/carbon catalyst, *J Molec Catal A Chem*, 2007, *265*, 80-89.

[31] a) C.S. Srikanth, V.P. Kumar, B. Viswanadham, K.V.R. Chary, Hydrodechlorination of 1,2,4-trichlorbenzene over supported ruthenium catalysts on various supports, *Catal Commun*, 2011, *13*, 69-72; b) T. Yoneda, T. Takido, K. Konuma, Hydrodechlorination of para-substituted

chlorobenzenes over a ruthenium/carbon catalyst, *Appl Catal B Environ*, 2008, *84*, 667-677.

[32] G.S. Pozan, I. Boz, Catalytic hydrodechlorination of 2,4-dichlorophenol on Pd/Rh/C catalysts, *J Haz Mat*, 2006, *136(3)*, 917-921; b) E. Díaz, J.A. Casas, A.F. Mohedano, L. Calvo, M.A. Gilarranz, J.J. Rodríguez, Kinetics of 4-chlorophenol hydrodechlorination with alumina and activated carbon-supported Pd and Rh catalysts, *Ind Eng Chem Res*, 2009, *48*, 3351-3358.

[33] M.A. Keane, P.M. Patterson, In situ production of nickel/graphitic carbon nanofiber composites and application in catalytic hydrodechlorination, *Catal Lett*, 2005, *99(1-2)*, 33-39; b) M.A. Keane, C.Park, C. Menini, Structure sensitivity in the hydrodechlorination of chlorobenzene over supported nickel, *Catal Lett*, 2003, 88(1-2), 89-94; c) Y. Wang, J. Wang, G. Fan, F. Li, Synthesis of a novel Ni/C catalyst derived from a composite precursor for hydrodechlorination, *Catal Commun*, 2012, *19*, 56-60; d) S. Gryglewicz, W. Piechocki, Hydrodechlorination of dichlorobenzenes and their derivatives over Ni–Mo/C catalyst: Kinetic analysis and effect of molecular structure of reactant, *Chemosphere*, 2011, *83*, 334-339; e) W. Piechocki, G. Gryglewicz, S. Gryglewicz, Hydrodechlorination of DDT and chloroalkanes over carbon-supported Ni–Mo catalyst, *J Haz Mat*, 2009, *163*, 1397-1402; f) J. Wang, G. Fan, F. Li, Carbon-supported Ni catalysts with enhanced metal dispersion and catalytic performance for hydrodechlorination of chlorobenzene, *RSC Advances*, 2012, *2*, 9976-9985; g) A. Srebowata, I. Stefanowicz-Pieta, W. Juszczyk, Z. Karpinski, Chlorine removal from 1,2-dichloroethane over Ni/C catalysts, *Polish J Chem*, 2007, *81*, 1521-1529.

[34] Y. Sun, M. Takaoka, N. Takeda, T. Matsumoto, K. Oshita, Kinetics on the decomposition of polychlorinated biphenyls with activated carbon-supported iron, *Chemosphere*, 2006, *65*, 183-189; b) Y. Sun, R. Nagao, M. Takaoka, N. Takeda, T. Matusmoto, K. Oshita, K., Catalytic degradation of polychlorinated biphenyls with activated carbon-supported iron, *J Environ Chem*, 2005, *15*, 259-268.

[35] E. Diaz, A.F. Mohedano, J.A. Casas, L. Calvo, M.A. Gilarranz, J.J. Rodriguez, Comparison of activated carbon-supported Pd and Rh catalysts for aqueous-phase hydrodechlorination, *Appl Catal B Environ*, 2011, *106*, 469-475.

[36] a) Y. Mitoma, M. Takase, Y. Yoshino, T. Masuda, H. Tashiro, N. Egashira, T. Oki, Novel mild hydrohydrodechlorination of PCDDs,

PCDFs, and co-PCBs inside fly ash using a calcium-promoted rhodium carbon catalyst in methanol, *Environ Chem*, 2006, *38*, 215-218; b) b) Y. Mitoma, N. Egashira, C. Simion, Highly effective degradation of polychlorinated biphenyls in soil mediated by a Ca/Rh bicatalytic system. *Chemosphere*, 2009, *74*, 968-973; c) Y. Mitoma, M. Kakeda, A.M. Simion, N. Egashira, C. Simion, Metallic Ca-Rh/C-methanol, a high-performing system for the hydrodechlorination/ring reduction of mono- and poly chlorinated aromatic substrates. *Environ Sci Technol*, 2009, *43(15)*, 5952-5958.

[37] M. Martin-Martinez, L.M. Gómez-Sainero, M.A. Alvarez-Montero, J. Bedia, J.J. Rodriguez, Comparison of different precious metals in activated carbon-supported catalysts for the gas-phase hydrodechlorination of chloromethanes, *Appl Catal B Environ*, 2013, *132-133*, 256-265.

[38] H.W. Kroto, J.R. Heath, S.C. O'Brien, R.F. Curl, R.E. Smalley, C_{60}: Buckminsterfullerene, *Nature*, 1985, *318*, 162-163.

[39] S. Iijima, Helical microtubules of graphitic carbon, *Nature (London)*, 1991, *354*, 56-58.

[40] a) R. Kuzuo, M. Terauchi, M. Tanaka, Y. Saito, H. Shinohara, Electron-energy-loss spectra of crystalline C_{84}, *Phys Rev B*, 1994, *49(7)*, 5054-5057; b) D.L. Dorset, J.R. Fryer, Quantitative electron crystallographic determinations of higher fullerenesin the hexagonal close packed polymorph, *J Phys Chem B*, 2001, *105(12)*, 2356-2359.

[41] M. Ge, K. Sattler, Observation of fullerene cones, *Chem Phys Lett*, 1994, *220*, 192-196.

[42] S. Itoh, S. Ihara, J. Kitakami, Toroidal form of carbon C360, *Phys Rev B*, 1993, *47*, 1703-1704.

[43] S. Amelinckx, X.B. Zhang, D. Bernaerts, X.F. Zhang, V. Ivanov, J.B. Nagy, A formation mechanism for catalytically grown helix-shaped graphite nanotubes, *Science*, 1994, *265*, 635-639.

[44] A.L. Mackay, H. Terrones, Diamond from graphite, *Nature*, 1991, *352*, 762.

[45] H. Terrones, M. Terrones, Curved nanostructured materials, *New J Phys*, 2003, *5*, 126.1–126.37.

[46] D.R. Dreyer, R.S. Ruoff, C.W. Bielawski, From conception to realization: an historial account of graphene and some perspectives for its future, *Angew Chem Int Ed*, 2010, *49*, 9336-9345.

[47] a) T.V. Reshetenko, L.B. Avdeeva, Z.R. Ismagilov, A.L. Chuvilin, Catalytic filamentous carbon as supports for nickel catalysts, *Carbon*,

2004, *42*, 143-148; b) T.G. Ros, A.J. van Dillen, J.W. Geus, D.C. Koningsberger, Surface oxidation of carbon nanofibres, *Chem Eur J*, 2002, *8*, 1151-1162.

[48] J. Kua, W.A. Goddard III, Oxidation of methanol on 2nd and 3rd row group. VIII. Transition metals: applications to direct methanol fuel cells, *J Am Chem Soc*, 1999, *121*, 10928-10941.

[49] P. Serp, M. Corrias, P. Kalck, Carbon nanotubes and nanofibers in catalysis, *Appl Catal A Gen*, 2003, *253*, 337-358.

[50] a) C. Amorim, X. Wang, M.A. Keane, Application of hydrodechlorination in environmental pollution control: comparison of the performance of supported and unsupported Pd and Ni catalysts, *Chin J Catal*, 2011, *32*, 746-755; b) C. Amorim, M.A. Keane, Effect of surface acid groups associated with amorphous and structured carbon on the catalytic hydrodechlorination of chlorobenzenes, *J Chem Technol Biotechnol*, 2008, *83*, 662-672; c) C. Amorim, M.A. Keane, Palladium supported on structured and nonstructured carbon: A consideration of Pd particle size and the nature of reactive hydrogen, *J Coll Interface Sci*, 2008, *322*, 196-208.

[51] a) R.F. Bueres, E. Asedegbega-Nieto, E. Diaz, S. Ordonez, F.V. Diez, Performance of carbon nanofibres, high surface area graphites, and activated carbons as supports of Pd-based hydrodechlorination catalysts, *Catal Today*, 2010, *150*, 16-21, and references cited therein; b) S. Ordonez, E. Díaz, R.F. Bueres, E. Asedegbega-Nieto, H. Sastre, Carbon nanofibre-supported palladium catalysts as model hydrodechlorination catalysts, *J Catal*, 2010, *272*, 158-168

[52] E. Lopez, S. Ordonez, F.V. Díez, Deactivation of a Pd/A12O$_3$ catalyst used in hydrodechlorination reactions: Infl uence of the nature of organochlorinated compound and hydrogen chloride, *Appl Catal B*, 2006, 62, 57-65.

[53] R.F. Bueres, E. Asedegbega-Nieto, E. Díaz, S. Ordonez, F.V. Díez, Preparation of carbon nanofibres supported palladium catalysts for hydrodechlorination reactions, *Catal Commun*, 2008, *9*, 2080-2084.

[54] E. Díaz, S. Ordonez, R.F. Bueres, E. Asedegbega-Nieto, H. Sastre, High-surface area graphites as supports for hydrodechlorination catalysts: Tuning support surface chemistry for an optimal performance, *Appl Catal B Environ*, 2010, *99*, 181-190.

[55] E.S. Lokteva, S.A. Kachevskii, A.O. Turakulova, E.V. Golubina, V.V. Lunin, A.E. Ermakov, M.A. Uimin, A.A. Mysik, The hydrodechlorination of chlorobenzene in the vapor phase in the presence

of metal–carbon nanocomposites based on Nickel, Palladium, and Iron, *Russ J Phys Chem A*, 2009, *83(8)*, 1300-1306.

[56] Z. Jin, C. Yu, X. Wang, Y. Wan, D. Li, G. Lu, Hydrodechlorination of chlorophenols at low temperature on a novel Pd catalyst, *Chem Commun*, 2009, 4438-4440.

[57] M. Bonarowska, Z. Karpinski, Carbon nanotubes as a catalyst support in hydrodechlorination of dichlorodifluoromethane on metal catalysts, *Polish J Chem*, 2009, *83*, 1821-1830.

[58] a) E.V. Golubina, E.S. Lokteva, A.G. Majouga, M.V. Lobanov, V.V. Lunin, Ultradispersed diamond as an excellent support for Pd and Au nanoparticle based catalysts for hydrodechlorination and CO oxidation, *Diamond Rel Mat*, 2011, *20*, 960-964; b) E.S. Lokteva, E.V. Golubina, S.A. Kachevskii, A.N. Kharlanov, A.V. Erokhin, V.V. Lunin, Ultradispersed diamond as a new carbon support for hydrodechlorination catalysts, *Kinet Catal*, 2011, *52(1)*, 145-155; c) S.A. Kachevskii, E.V. Golubina, E.S. Lokteva, V.V. Lunin, Palladium on ultradisperse diamond and activated carbon: the relation between structure and activity in hydrodechlorination, *Russ J Phys Chem A,* 2007, *81(6)*, 866-873.

[59] A. Srebowata, W. Juszczyk, Z. Kaszkur, J.W. Sobczak, L. Kepinski, Z. Karpinski, Hydrodechlorination of 1,2-dichloroethane and dichlorodifluoromethane over Ni/C catalysts: the effect of catalyst carbiding, *Appl Catal A Gen*, 2007, *319*, 181-192.

[60] Q. Liu, Z.-M. Cui, Z. Ma, S.-W. Bian, W.-G. Song, Highly active and stable material for catalytic hydrodechlorination using ammonia-treated carbon nanofibers as Pd supports, *J Phys Chem C*, 2008, *112,* 1199-1203.

[61] J.A. Diaz, R. Diaz-Moreno, L.S. Silva, F. Dorado, A. Romero, J.L. Valverde, Nickel supported carbon nanofibers as an active and selective catalyst for the gas-phase hydrogenation of 2-*tert*-butylphenol, *J Coll Interface Sci*, 2012, 380, 173-181.

[62] O.V. Turova, E.V. Starodubtseva, M.G. Vinogradov, V.I, Sokolov, N.V. Abramova, A.Ya. Vul, A.E. Alexenskiy, Palladium supported on detonation nanodiamond as a highly effective catalyst of the C=C and C≡C bond hydrogenation, *Catal Commun*, 2011, *12*, 577-579.

[63] I. Pri-Bar, B.R. James, Mechanochemical, solvent free, palladium-catalyzed hydrodechlorination of chloroaromatic hydrocarbons, *J Molec Catal A Chem*, 2007, *264*, 135-139.

[64] See for example: a) O.Y. Posudievsky, O.A. Khazieieva, V.G. Koshechko, V.D. Pokhodenko, Preparation of graphene oxide by solvent-free mechanochemical oxidation of graphite, *J Mater Chem*, 2012, *22*, 12465-12467; b) O. Senneca, P. Salatino, R. Chirone, L. Cortese, R. Solimene, Mechanochemical activation of high-carbon fly ash for enhanced carbon reburning, *Proc Combustion Instit*, 2011, *33(2)*, 2743-2753; c) B. Todorović Marković, V. Jokanović, S. Jovanović, D. Kleut, M. Dramićanin, Z. Marković, Surface chemical modification of fullerene by mechanochemical treatment, *Appl Surface Sci*, 2009, *255(17)*, 7537-7541; d) D.V. Onishchenko, V.P. Reva, V.G. Kuryavyi, V.V. Petrov, Processing of sphagnum moss to produce carbon nanotubes, *Coke Chem*, 2012, *55(9)*, 358-361.

[65] N. Job, B. Heinrichs, S. Lambert, J.-P. Pirard, J.-F. Colomer, B. Vertruyen, J. Marien, Carbon xerogels as catalyst supports: study of mass transfer, *AIChE J*, 2006, *52(8)*, 2663-2676.

[66] a) U. Quaß, M. Fermann, G. Bröker, The European Dioxin Air Emission Inventory Project—Final Results, *Chemosphere*, 2004, *54(9)*, 1319-1327; b) E. Finocchio, G. Busca, M. Notaro, A review of catalytic processes for the destruction of PCDD and PCDF from waste gases, *Appl Catal B Environ*, 2006, 62, 12-20.

[67] K.H. Chi, S.H. Chang, C.H. Huang, H.C. Huang, M.B. Chang, Partitioning and removal of dioxin-like congeners in flue gases treated with activated carbon adsorption, *Chemosphere*, 2006, *64*, 1489-1498.

[68] a) L.C. Haws, S.H. Su, M. Harris, M.J. DeVito, N.J. Walker, W.H. Farland, B. Finley, L.S. Birnbaum, Development of a refined database of mammalian relative potency estimates for dioxin-like compounds, *Toxicol Sci*, 2006, *89*, 4-30; b) M. van den Berg, L.S. Birnbaum, M. Denison, M. DeVito, W. Farland, M. Feeley, H. Fiedler, H. Hakansson, A. Hanberg, L. Haws, M. Rose, S. Safe, D. Schrenk, C. Tohyama, A. Tritscher, J. Tuomisto, M. Tysklind, N. Walker, R.E. Peterson, The 2005 World Health Organization re-evaluation of human and mammalian toxic equivalency factors for dioxins and dioxin-like compounds, *Toxicol Sci*, 2006, *93*, 223-241.

[69] A.N. Tugov, V.R. Moskvichev, L.G. Fedorov, N.S. Sokolov, Methods for reducing emissions of dioxins and furans in flue gases at plants burning solid domestic waste, *Power Technol Eng*, 2009, *43(1)*, 46-51.

[70] Y.-M. Chang, C.-Y. Hung, J.-H. Chen, C.-T. Chang, C.-H. Chen, Minimum feeding rate of activated carbon to control dioxin emissions

from a large-scale municipal solid waste incinerator, *J Haz Mat*, 2009, *161*, 1436-1443.

[71] P.C. Hung, W.C. Lo, K.H. Chi, S.H. Chang, M.B. Chang, Reduction of dioxin emission by a multi-layer reactor with bead-shaped activated carbon in simulated gas stream and real flue gas of a sinter plant, *Chemosphere*, 2011, *82*, 72-77.

[72] A. Kawashima, M. Katayama, N. Matsumoto, K. Honda, Physicochemical characteristics of carbonaceous adsorbent for dioxin-like polychlorinated biphenyl adsorption, *Chemosphere*, 2011, *83*, 823-830.

[73] E. Díaz, S. Ordonez, A. Vega, Adsorption of volatile organic compounds onto carbon nanotubes, carbon nanofibers, and high-surface-area graphites, *J Coll Interface Sci*, 2007, *305*, 7-16.

[74] Y. Hajizadeh, J.A. Onwudili, P.T. Williams, Removal potential of toxic 2378-substituted PCDD/F from incinerator flue gases by waste-derived activated carbons, *Waste Manag*, 2011, *31*, 1194-1201.

[75] L. Zhao, H. Hou, K. Iwasaki, A. Terada, M. Hosomi, Utilization of recycled charcoal as a thermal source and adsorbent for the treatment of PCDD/Fs contaminated sediment, *J Haz Mat*, 2012, *225-226*, 182-189.

[76] J.L. Bonte, K.J. Fritsky, M.A. Plinke, M. Wilken, Catalytic destruction of PCDD/F in a fabric fllter: experience at a municipal waste incinerator in Belgium, *Waste Manag*, 2002, *22*, 421-426.

[77] A. Elola, E. Diaz, S. Ordonez, A new procedure for the treatment of organochlorinated off-gases combining adsorption and catalytic hydrodechlorination, *Environ Sci Technol*, 2009, *43,* 1999-2004.

[78] H. Choi, S.R. Al-Abed, S. Agarwal, Catalytic role of Palladium and relative reactivity of substituted chlorines during adsorption and treatment of PCBs on reactive activated carbon, *Environ Sci Technol*, 2009, *43,* 7510-7515.

[79] S.H. Chang, J.W. Yeh, H.M. Chein, L.Y. Hsu, K.H. Chi, M.B. Chang, PCDD/F Adsorption and destruction in the flue gas streams of MWI and MSP via Cu and Fe catalysts supported on carbon, *Environ Sci Technol*, 2008, *42,* 5727-5733.

[80] Johannes Lehmann, Stephen Joseph in "*Biochar for Environmental Management: Science and Technology*", Earthscan, Dunstan House, London, 2009.

[81] S.E. Hale, J. Lehmann, D. Rutherford, A.R. Zimmerman, R.T. Bachmann, V. Shitumbanuma, A. O'Toole, K.L. Sundqvist, H.P.H. Arp, G. Cornelissen, Quantifying the total and bioavailable polycyclic

aromatic hydrocarbons and dioxins in biochars, *Environ Sci Technol*, 2012, *46*, 2830-2838.

[82] Y. Chai, R.J. Currie, J.W. Davis, M. Wilken, G.D. Martin, V.N. Fishman, U. Ghosh, Effectiveness of activated carbon and biochar in reducing the availability of polychlorinated dibenzo-*p*-dioxins/dibenzofurans in soils, *Environ Sci Technol*, 2012, *46*, 1035-1043.

[83] a) Y. Chai, J.W. Davis, M. Wilken, G.D. Martin, D.M. Mowery, U. Ghosh, Role of black carbon in the distribution of polychlorinated dibenzo-p-dioxins/dibenzofurans in aged field-contaminated soils, *Chemosphere*, 2011, *82*, 639-647; b) H.-C. Hsi, L.-C. Wang, T.-H. Yu, Effects of injected activated carbon and solidification treatment on the leachability of polychlorinated dibenzo-p-dioxins and dibenzofurans from air pollution control residues of municipal waste incineration, *Chemosphere*, 2007, *67*, 1394-1402.

[84] E. Sanchez-Cortezon, "The role of lattice defects in the catalytic oxidation of methanol over graphite catalysts", Ph.D. thesis, Technische Universität, Berlin, 2002.

[85] a) M.F.R. Pereira, J.L. Figueiredo, J.J.M. Orfao, P. Serp, P. Kalck, Y. Kihn, Catalytic activity of carbon nanotubes in the oxidative dehydrogenation of ethylbenzene, *Carbon*, 2004, *42(14)*, 2807-2813; b) J.J. Delgado, X. Chen, J.P. Tessonnier, M.E. Schuster, E. Del Rio, R. Schlögl, D.S. Su, Influence of the microstructure of carbon nanotubes on the oxidative dehydrogenation of ethylbenzene to styrene, *Catal Today*, 2010, *150(1-2)*, 49-54.

[86] J.J. Delgado, X.-W. Chen, B. Frank, D.S. Su, R. Schlogl, Activation processes of highly ordered carbon nanofibers in the oxidative dehydrogenation of ethylbenzene, *Catal Today*, 2012, *186(1)*, 93-98.

[87] J. Zhang, D.S. Su, R. Blume, R. Schlogl, R. Wang, X. Yang, A. Gajovic, Surface chemistry and catalytic reactivity of a nanodiamond in the steam-free dehydrogenation of ethylbenzene, *Angew Chem Int Ed*, 2010, *49*, 8640-8644.

[88] a) X. Liu, B. Frank, W. Zhang, T.P. Cotter, R. Schlogl, D.S. Su, Carbon-catalyzed oxidative dehydrogenation of n-butane: selective site formation during sp^3-to-sp^2 lattice rearrangement, *Angew Chem Int Ed*, 2011, *50*, 3318-3322; b) J. Zhang, X. Liu, R. Blume, A. Zhang, R. Schlogl, D.S. Su, Surface-modified carbon nanotubes catalyze oxidative dehydrogenation of *n*-butane, *Science*, 2008, *322*, 73-77.

[89] D.Y. Jang, H.G. Jang, G.R. Kim, G.-J. Kim, Oxidative dehydrogenation of n-butane on nano-carbon catalysts having graphitic structures, *Res Chem Intermed*, 2011, *37*, 1145-1156.

[90] Z. Sui, J. Zhou, Y. Dai, W. Yuan, Oxidative dehydrogenation of propane over catalysts based on carbon nanofibers, *Catal Today*, 2005, *106*, 90-94.

[91] D. Begin, G. Ulrich, J. Amadou, D.S. Su, C. Pham-Huu, R. Ziessel, Oxidative dehydrogenation of 9,10-dihydroanthracene using multi-walled carbon nanotubes, *J Mol Catal A Chem*, 2009, *302*, 119-123.

[92] Kh.M. Alimardanov, A.A. Alieva, S.I. Abasov, M.F. Abbasov, A.D. Kuliev, Effect of immobilized carbon nanoparticles on the activity of zeolites in the oxidative dehydrogenation of 4-vinylcyclohexene and ethylbenzene to styrene, *Petroleum Chem*, 2012, *52(2)*, 97-104.

[93] E.A. Tveritinova, I.I. Kulakova, Yu.N. Zhitnev, A.V. Fionov, A. Lund, W. Chen, I. Buyanova, V.V. Lunin, Catalytic conversion of C(2)-C(3) alcohols on detonation nanodiamond and its modifications, *Russ J Phys Chem*, 2012, *86(1)*, 26-31.

[94] D.R. Dreyer, H.-P. Jia, C.W. Bielawski, Graphene oxide: a convenient carbocatalyst for facilitating oxidation and hydration reactions, *Angew Chem Int Ed*, 2010, *49*, 6813-6816.

[95] J. Zhang, D.S. Su, A. Zhang, D. Wang, R. Schlogl, C. Hebert, Nanocarbon as robust catalyst: mechanistic insight into carbon-mediated synthesis, *Angew Chem Int Ed*, 2007, *46*, 7319-7323.

[96] G. Mestl, N.I. Maksimova, N. Keller, V.V. Roddatis, R. Schlogl, Carbon nanofilaments in heterogeneous catalysis: an industrial application for new carbon materials?, *Angew Chem Int Ed*, 2001, *40*, 2066-2068.

[97] a) D.R. Dreyer, S. Park, C.W. Bielawski, R.S. Ruoff, The chemistry of graphene oxide, *Chem Soc Rev*, 2010, *39*, 228-240; b) K.P. Loh, Q.L. Bao, G. Eda, M. Chhowalla, M. Graphene oxide as a chemically tunable platform for optical applications, *Nat Chem*, 2010, *2*, 1015-1024.

[98] A. Rinaldi, J. Zhang, B. Frank, D.S. Su, S.B.A. Hamid, R. Schlogl, Oxidative purification of carbon nanotubes and its impact on catalytic performance in oxidative dehydrogenation reactions, *ChemSusChem*, 2010, *3*, 254-260.

[99] a) Y. Wang, X.C. Wang, M. Antonietti, Polymeric graphitic carbon nitride as a heterogeneous organocatalyst: from photochemistry to multipurpose catalysis to sustainable chemistry, *Angew Chem Int Ed*, 2012, *51*, 68-89; b) D.S. Su, J. Zhang, B. Frank, A. Thomas, X.C. Wang, J. Paraknowitsch, R. Schlogl, Metal-free heterogeneous catalysis for

sustainable chemistry, *ChemSusChem*, 2010, *3*, 169-180; c) J.H. Bitter, Nanostructured carbons in catalysis a janus material-industrial applicability and fundamental insights, *J Mater Chem*, 2010, *20*, 7312-7321; d) D.R. Dreyer, C.W. Bielawski, Carbocatalysis: heterogeneous carbons finding utility in synthetic chemistry, *Chem Sci*, 2011, *2*, 1233-1240.

[100] C. Su, K.P. Loh, Carbocatalysts: graphene oxide and its derivatives, *Acc Chem Res*, 2012, Article ASAP, DOI: 10.1021/ar300118v

[101] a) Y. Kakuta, T. Matsuto, N. Tanaka, T. Masuda, Influence of residual carbon on the decomposition process PCDD/Fs in MSWI fly ash, *Chemosphere*, 2005, *58*, 969-975.; b) Y. Kakuta, T. Matsuto, Y. Tojo, H. Tomikawa, Characterization of residual carbon influencing on de novo synthesis of PCDD/Fs in MSWI fly ash, *Chemosphere*, 2007, 68, 880-886.

[102] M. Farcasiu, P.B. Kaufman, E.P. Ladner, R.R. Anderson, Method of dehalogenation using diamonds, US Pat 6,143,939, 2000.

[103] a) A. Ghaffar, M. Tabata, Optimized catalytic dechlorination of chlorobenzene compounds on fly ash; role of metals, aqueous/organic solvents and temperatures, *Appl Catal B Environ*, 2009, *86*, 152-158; b) A. Ghaffar, M. Tabata, Dechlorination/detoxification of aromatic chlorides using fly ash under mild conditions, *Waste Manag*, 2009, 29, 3004-3008; c) A. Ghaffar, M. Tabata, Dechlorination of p-nitrochlorobenzene by using the fly ash; effects of metals; mixed solvents and temperatures, *Reac Kinet Catal Lett*, 2009, 97(2), 35-41; c) A. Ghaffar, M. Tabata, Enhanced dechlorination of chlorobenzene compounds on fly ash: effects of metals, solvents and temperature, *Green Chem Lett Rev*, 2010, *3(2)*, 179-190.

[104] A. Ghaffar, M Tabata, F. Mazher, A. Mashiatullah, High dechlorination of PCBs catalysed by carbon under mild conditions, *Environ Chem Lett*, 2012, *10*, 171-176.

[105] D. Yu, E. Nagelli, F. Du, L. Dai, Metal-free carbon nanomaterials become more active than metal catalysts and last longer, *J Phys Chem Lett*, 2010, *1*, 2165-2173.

In: Charcoal
Editor: Eric N. Tyrone
ISBN: 978-1-62808-663-8

Chapter 2

EMISSION REDUCTION THEORY AND RESULTS IN THE DEVELOPMENT OF A SUITABLE SMALL-SCALE, PORTABLE, CONTINUOUS RICE HUSK CARBONISER FOR RURAL REGIONS IN TRANSITIONAL ECONOMIES

***Ricardo F. Orge*[1,*] *and Mark P. McHenry*[2,†]**
[1]Philippine Rice Research Institute (PhilRice),
Maligaya, Muñoz Science City, Nueva Ecija, Philippines
[2]School of Engineering and Information Technology,
90 South St. Murdoch University, Murdoch WA, Australia

ABSTRACT

This chapter describes the process of developing a new rice husk conversion technology suitable to the local requirements of rice farmers in poor rural areas of the Philippines. The small-scale, portable,

* Corresponding author: Ricardo F. Orge. Philippine Rice Research Institute (PhilRice), Maligaya, Muñoz Science City, 3119 Nueva Ecija, Philippines. Tel: (+63) (44) 456-0258; E-mail: rforge@gmail.com.

† Mark P. McHenry: School of Engineering and Information Technology, 90 South St. Murdoch University, Western Australia, 6150.

continuous carboniser was designed to enable heat extraction applications (crop drying, steam generation, etc.), while maintaining conversion effectiveness, with a special focus on the reduction of rural workplace carbon monoxide and smoke emissions. The chapter contains the technology optimization model methodology which aimed to achieve high operating temperatures for heat extraction, highest fixed carbon content of rice husk biochar, and the lowest emissions of carbon monoxide possible during the carbonisation process. A Response Surface Methodology was used to develop predictive models relating carboniser performance parameters (airflow rate, temperature, biochar quality, and carbon monoxide emission data) to the carboniser parameters (chimney diameter and ignition chamber opening). The analysis of variance showed the importance of the chimney diameter relative to the ignition chamber opening in determining overall performance. The high temperatures and low CO emissions were associated with the high airflow rates, although proximate analysis of the biochar samples showed there was no influence of the relative sizes of the chimney and ignition chamber on the biochar fixed carbon content. Therefore, the selection of the optimal carboniser chimney and ignition chamber sizes were based solely on temperature and CO emissions.

Keywords: Carboniser; biochar; bioenergy; rice; rice husk; emissions; carbon monoxide

NOMENCLATURE

A	area of the air inlets
RSM	response surface methodology
D	overall desirability considering all response factor in the optimisation
d_i	transformation of response variable i into a desirability function
r_i	the importance factor for response variable i, and
n	number of factors being optimised
IC	ignition chamber
V_b	air velocity at the bottom of the carboniser (m s^{-1})
V_s	air velocity at the side opening of the carboniser (m s^{-1})
Q	airflow rate (m^3 h^{-1})
h_{ic}	height of the ignition chamber opening (cm)
d_c	the diameter of the chimney (cm)

INTRODUCTION

Rice husks, the outer covering of rice grains removed during milling, constitute around 20 to 25% of the total grain harvest mass and are thus an abundant agricultural residue [1]. This chapter explores the theoretical development and actual testing of a constructed small-scale, non-motorised, continuous rice husk carboniser developed by the Philippine Rice Research Institute (PhilRice). The prototype was able to process 40 kg of rice husk into biochar each hour at 40.3% biochar yield (by weight), with a purity of 99.1% biochar (see Figure 1). Unlike existing batch rice husk carbonisers in common use, the prototype produced visibly clean emissions from the chimney during operation. However, as carboniser operators in poor transitional countries are often exposed to excessive smoke and invisible carbon monoxide (CO) emissions as part of their workplace, this chapter records a methodology for technology developments of optimised bioenergy systems that aim to reduce operator exposures, in this case for carbonisation of rice husks in poor rural rice producing regions.

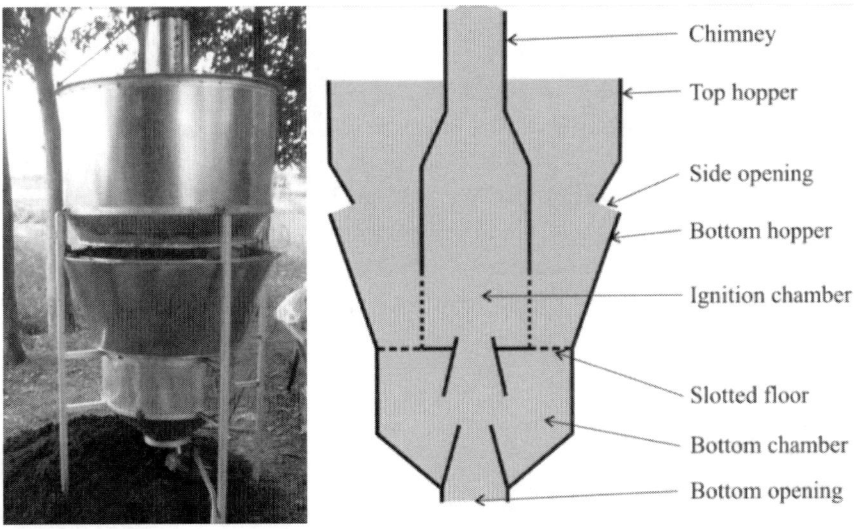

Figure 1. The prototype of the Phil Rice continuous rice husk carboniser used in the study.

Carboniser performance assessments often require mathematical models to approximate the complex behavior of bioenergy systems [2, 3]. This chapter describes the development of a mathematical model for optimizing rice husk carboniser performance in terms of maximizing operating temperatures, minimizing the ash content of the output biochar, and also CO emissions.

While developing an "optimal design" is impossible in practice due to the complexity in system characteristics uncertainties [4], nonetheless, carboniser performance can be evaluated in terms of temperature consistency, biochar conversion efficiency, and by-product emissions [5].

MATERIALS AND METHODS

Model Formulation

A portable, continuous, rice husk carboniser prototype with provisions for varying intake air volumes was developed to generate primary input data for the model. The performance parameters were modelled as a function of carboniser parameters, based on the assumption that the following general relationships exist:

- Airflow rate = f(carboniser parameters)
- Temperature = f(airflow rate)
- Biochar (quality) = f(temperature)
- Emission = f(airflow rate)

Several carboniser parameters were identified to have varying influence on the amount of air taking part in the carbonisation process (Table 1).

However, the final independent variables selected in the model were chimney diameter, and area of the ignition chamber (IC) opening, as together they had the major influence on airflows.

Chimney Diameter

Among the two chimney parameters (height and diameter), the diameter was selected as the independent variable as it directly influenced the airflow rate in and out of the system.

Table 1. Initial carboniser parameters considered as independent variables

Component	Possible independent variable	Perceived influence on airflow/ temp.
Chimney	Diameter	Rate of flow of primary and secondary air
	Height	Static head air movement (natural draft)
Ignition chamber (IC)	Diameter	Area of heated surface
	Height	Area of heated surface
	Area of side opening	Rate of flow of partial combustion products
Bottom chamber	Diameter of bottom opening	Rate of flow of secondary air
Side opening	Area of rice husk bed surface	Rate of flow of primary air; flow resistance
Main hopper	Length of span of rice hull bed	Resistance to flow of primary air

Figure 2. The three removable carboniser chimneys used in the experiment.

A detachable chimney allowed three chimney diameters to be trialled: 7.5, 10, and 12.5 cm, corresponding to a cross-sectional area of 45.6, 81.1, and 126.7 cm^2, respectively (see Figure 2). The chimney height, which determines the static head causing airflow and overcoming resistance of the porous bed of rice husk, was fixed at 2.8 m for all the three chimney diameters. Design calculations in determining the chimney length for a bed span of 34.5 cm made use of the Ergun equation [6], following the same data and assumptions in Table 2.

Table 2. Data and assumptions used to calculate the chimney height

Parameters	Symbol	Value	Unit	Source
Void fraction (porosity)	ε	0.64	--	Ramirez *et al.* [7]
Particle sphericity	Φ	0.60	--	Ramirez *et al.* [7]
Equivalent diameter of particle	d_p	0.0016	m	Fang *et al.* [8]
Density of ambient air	ρ_1	1.16	kg m^{-3}	Lide [9]
Density of inside air (flue gas)	ρ_2	0.67	kg m^{-3}	Lide [9]
Air dynamic viscosity at 30°C	μ	0.00002	Pa sec^{-1}	Denny [10]

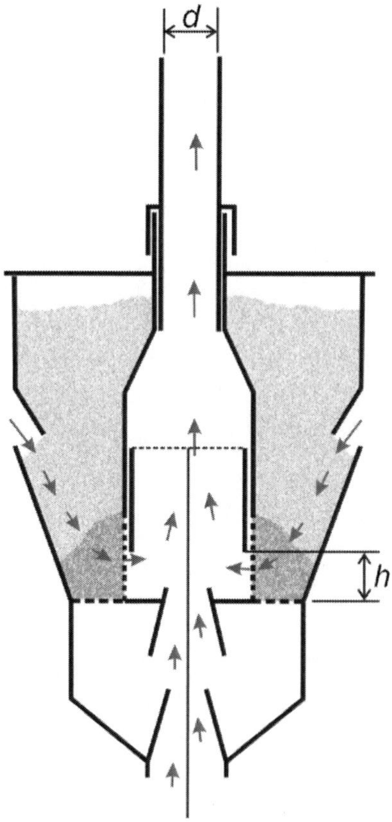

Figure 3. Schematic diagram showing the two carboniser parameters, chimney diameters (*d*), and IC areas (*h*) which were altered during the modelling experiment.

Area of the IC Opening

The openings at the side of the IC serve as passageway for the gaseous products from the reduction zone, enabling them to enter into the IC and mix with air for combustion prior to emission from the chimney.

It was assumed that varying this area would influence the equivalence ratio, and correspondingly the operating temperature [6].

Therefore, a hollow circular cylinder was designed with an outside diameter sufficient to freely slide inside the IC to enable altering the area of the IC opening (Figure 3). Three levels were used: 434, 868, and 1,302 cm^2 corresponding to a height, *h*, of 6, 12, and 18 cm, respectively.

In addition to the ambient air temperature, relative humidity, chimney temperature, and rice hull elemental data (ultimate analysis), for each combination of carboniser parameters (independent variables), the following performance parameters (dependent variables) were determined: airflow rates at the entry points; temperature at the IC; quality of biochar output, and; gas emission data.

Airflow Rates at the Entry Points

The air velocity at the bottom (V_b) and at the side openings (V_s) were measured using the VelociCheck™ TSI Model 8330-M (Figure 4) having a least count of 0.01 m s^{-1}. To facilitate measurement of the air velocity at the bottom opening, a 20 cm diameter elbow made of galvanised iron sheet was attached to the bottom opening, as shown in Figure 4. Two 8 mm holes spaced at 90 degrees were drilled 30 cm from the free end of the tube where the instrument's probe was inserted. Airflow measurement at the bottom opening was taken as an average of six readings, three taken from each hole. For the side opening, airflow measurements were obtained by positioning the instrument's probe 1 cm from the rice hull bed surface (Figure 5). Readings were taken at four locations approximately spaced 90 degrees from each other. To ensure that the air velocity readings were unaffected by prevailing winds, a 1.8 m height perimeter fence covered with orange laminated plastic sheet was installed at the test site. The velocity readings (m s^{-1}) were converted into airflow ($m^3 h^{-1}$) data (see Equation 1):

$$Q = A \times V \times 3600 \quad (1)$$

where: Q = airflow rate, $m^3 h^{-1}$; A = area of the air inlets, 0.2137 and 0.0314 m^2 for the side and bottom openings, respectively, and; V = air velocity, $m s^{-1}$.

Temperature at the IC

The temperature at the IC was determined by two 18 mm diameter and 0.5 m long ceramic lined thermocouple probes attached to a Yokogawa Portable Multi-Thermometer, Model 2423A, data recorder using type K thermocouple wires (see Figure 6).

Figure 4. Measuring the airflow at the bottom opening using the using the VelociCheck™ TSI Model 8330-M airflow meter.

Figure 5. Measuring airflow at the side opening of the carboniser using the VelociCheck™ TSI Model 8330-M.

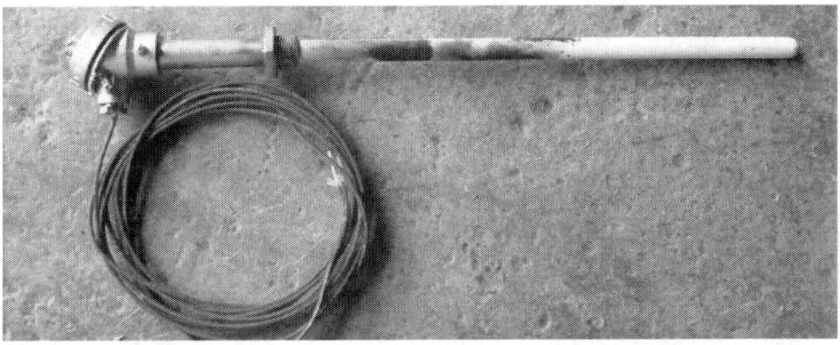

Figure 6. The temperature probes with type K thermocouple wires.

These were installed at the top and bottom portion of the IC with their tips approximately at the longitudinal axis of the IC (see Figure 7). Five temperature readings, spaced at five minutes, were taken from each point, and the final temperature data in this chapter were an average of these readings.

Quality of Biochar Output

Biochar quality was measured in terms of the fixed carbon content of the collected sample, expressed in a percent weight basis.

Approximately 500 g samples of the biochar were collected from each combination of independent parameters, and were submitted to the Analytical Service Laboratory of PhilRice for proximate analysis.

Gas Emission Data

The O_2, CO, and CO_2 content of the exhaust gas were determined using an IMR® 1000 handheld gas analyser (Figure 8). These gases were taken as a measure of combustion efficiency [11]. For example, if CO is high, there is deficiency of air which means that the combustion is absent or incomplete.

EXPERIMENTAL DESIGN AND ANALYSIS

Response surface methodology (RSM) was used to develop predictive models to determine how the carboniser performance parameters (airflow rate, temperature, biochar quality, and emission data) relate to the carboniser parameters (chimney diameter and IC opening), and were used to optimise responses affected by quantitative factors [12].

Figure 7. The positions of the thermocouples in the carboniser during data collection.

Figure 8. Gas emission data collection using the IMR® 1000 handheld gas analyser.

The modelling experiment was carried out following a face centred type of Central Composite Design of RSM [13], with the two carboniser parameters as independent variables each at three levels (see Figure 9).

The Design Expert® Statistical Software was employed to analyse the response data, develop predictive models, perform the various statistical computations, including tests for significance of the regression models and on individual model coefficients, lack-of-fit test, and a numerical optimisation of the carboniser parameters. The complete experimental design consisted of 14 experiments with four factorial experiments (levels -1 and +1), four axial experiments (levels +a and –a), and six replicated experiments for the centre of the design. The six replicated experiments at the centre were performed to facilitate estimation of pure error [14].

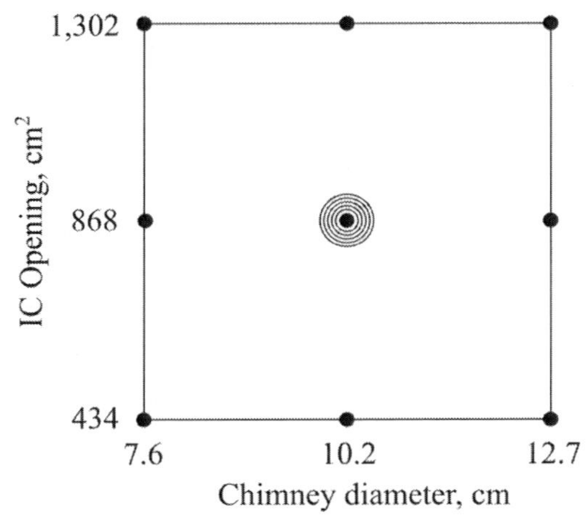

Figure 9. Face centred central composite design for the two independent parameters used.

The Experimental Model

To determine the data required to establish a mathematical model, the carboniser prototype was operated at all combinations of the independent parameters (chimney diameter and IC opening), (see Figure 10 for a flow chart of the experiment). The rice husks used in the experiment were derived from a rubber roll-type rice mill.

To ensure homogeneity, the rice hulls were mixed thoroughly and dried to a moisture content of 7.8% prior to the experiment. See Table 3 for additional husk data used in the material and heat balance analysis. After igniting, the carboniser was allowed to operate until the pyrolysed zone had reached the rice husk bed surface at the side opening, as visualised by the appearance of dark (burned) areas at the side opening. With the appearance of the dark areas, agitating the rice husk bed was done and gathering of data on the various dependent parameters followed immediately after.

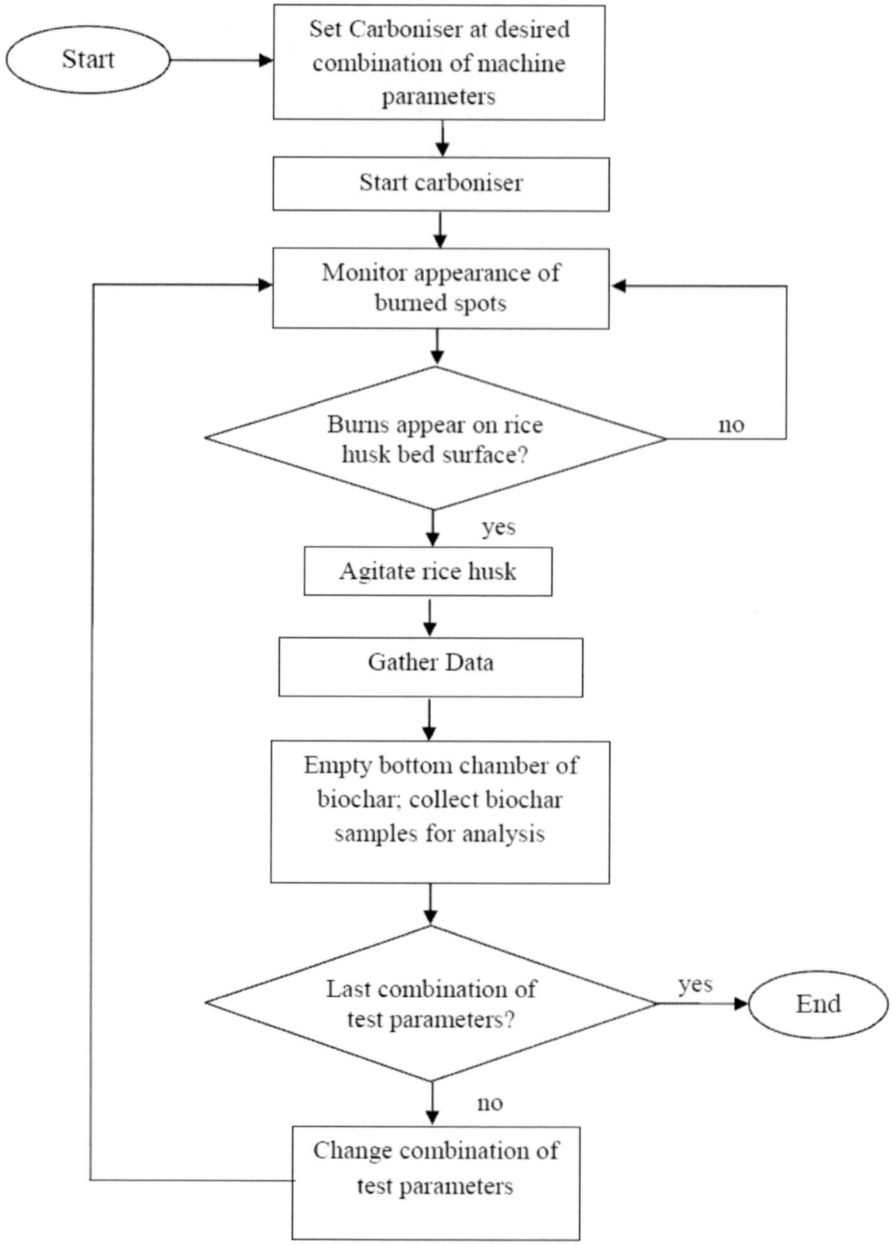

Figure 10. The experimental process expressed diagrammatically.

Table 3. Rice husk data used in the material and heat balance analysis

Parameter	Value
Proximate Analysis	
Moisture, (%)	7.8
Ash, (%)	18.1
Volatile, (%)	55.5
Fixed Carbon, (%)	18.6
Ultimate Analysis	
Carbon, (%)	38.2
Hydrogen, (%)	5.5
Oxygen, (%)	35.2
Nitrogen, (%)	0.9
Ash, (%)	20.2

Waiting for the pyrolysed zone to occur and agitating the rice husk bed before starting to collect the data were undertaken to minimise, if not eliminate, the experimental bias for data gathered between the first combination of the independent parameters (when initial ignition was undertaken) and the succeeding combinations of the independent parameters. During the tests, the hopper was always maintained at its full capacity.

RESULTS AND DISCUSSION

Effects of Carboniser Parameters on Carboniser Performance

Summary data on carboniser performance at different combinations of independent parameters are shown in Table 4 and 5. In general, combustion of the gases at the IC occurred in all combinations of the independent parameters, as shown by the visual presence of flames inside the IC. This however excluded experiments with the smallest chimney (7.5 cm diameter) in combination with the smallest opening of the IC (6 cm), which was the only combination that resulted in smoke emissions from the chimney during the conduct of the experiment.

The test for significance of the regression models, on individual model coefficients, and lack-of-fit tests were performed using the Design Expert®

Statistical software. Selecting the step-wise regression method eliminated the insignificant model terms, and the ANOVA tables (see Appendices 1 to 6) are presented for each dependent parameter.

The Appendices also show the significant model terms as well as the R^2, adjusted R^2, and predicted R^2. Except for the biochar fixed carbon content, the adequacy measures of the performance parameters approximate 1, which indicates model adequacy.

The sufficient precision values compare the range of the predicted values at the design points to the average prediction error. In all cases, the values of adequate precision were dramatically greater than 4. The adequate precision value above 4 indicate model discrimination adequacy, meaning that the model was suitable to navigate the design space.

Table 4. Summary data taken during optimisation test runs

Run	Chimney diameter (cm)	IC opening,* (cm)	Total Airflow** (m^3/h)	Temp*** (°C)	biochar quality (% fixed carbon)	Gas Emission		
						O_2 (%)	CO (ppm)	CO_2 (%)
1	7.5	6	50.19	417.6	13.3	18.7	1345	0.0
2	10	12	132.96	731.6	10.3	15.2	1140	3.0
3	12.5	6	204.42	896.8	15.2	8.0	530	7.2
4	12.5	18	224.36	909.8	13.7	6.0	450	5.1
5	10	12	130.84	735.6	10.3	16.0	1204	3.8
6	10	12	128.67	740.0	17.0	16.5	1238	3.1
7	7.5	18	61.08	571.2	11.4	15.5	1262	3.0
8	10	12	129.66	725.5	12.4	15.0	1126	2.7
9	10	18	123.84	742.8	16.3	14.6	852	4.4
10	10	12	135.18	742.9	14.2	15.8	1185	3.1
11	12.5	12	201.45	885.7	23.3	7.2	540	7.0
12	10	6	114.09	614.2	15.0	15.4	1159	3.0
13	10	12	128.22	780.3	10.3	15.2	1140	3.8
14	7.5	12	49.62	486.4	9.8	18.1	1358	1.6

*Areas of opening: 434, 868, and 1,302 cm^2 for heights of 6, 12, and 18 cm, respectively.

**Sum of the airflows at the bottom and side openings.

***Average of two temperature readings (taken at the top and bottom part of the IC).

Table 5. Experimental temperatures obtained at the top and bottom of the IC

Chimney diameter (cm)	IC opening (cm)	Temperature, °C		
		Top	Bottom	Average
7.5	6	476.2	359.0	417.6
7.5	12	549.8	423.0	486.4
7.5	18	645.3	497.1	571.2
10	6	682.1	546.3	614.2
10	12	820.4	659.7	740.1
10	12	815.3	647.9	731.6
10	12	805.8	645.1	725.5
10	12	867.8	692.8	780.3
10	12	824.8	661.0	742.9
10	12	818.5	652.7	735.6
10	18	821.6	664.1	742.9
12.5	6	952.3	841.3	896.8
12.5	12	948.2	823.2	885.7
12.5	18	965.7	853.9	909.8

Airflow Results

The effect of the chimney diameter and the size of opening of the IC on airflow is shown in Figure 11. Each data point was obtained by the sum of the airflow measurements from the bottom and side openings of the carboniser. The results show that both carboniser parameters influence the amount of air that took part in the carbonisation process. The effect of the chimney diameter, however, is more significant than that of the IC opening, as shown in the analysis of variance (see Appendix 1), as the effect of the chimney diameter is highly significant. The results are represented mathematically in Equation 2.

$$\text{Airflow} = -196.802 + 31.289 \times d_c + 1.127 \times h_{ic} \tag{2}$$

The model shows that as the diameter of the chimney (d_c) is increased, there is a corresponding linear increase in the amount of air entering into the carboniser, and a similar trend was observed for the size of the opening of the IC. As would be expected, the highest airflow rate was observed at the combination of the largest chimney diameter (12.5 cm) and the largest opening of the IC (18 cm) (see Appendix 1).

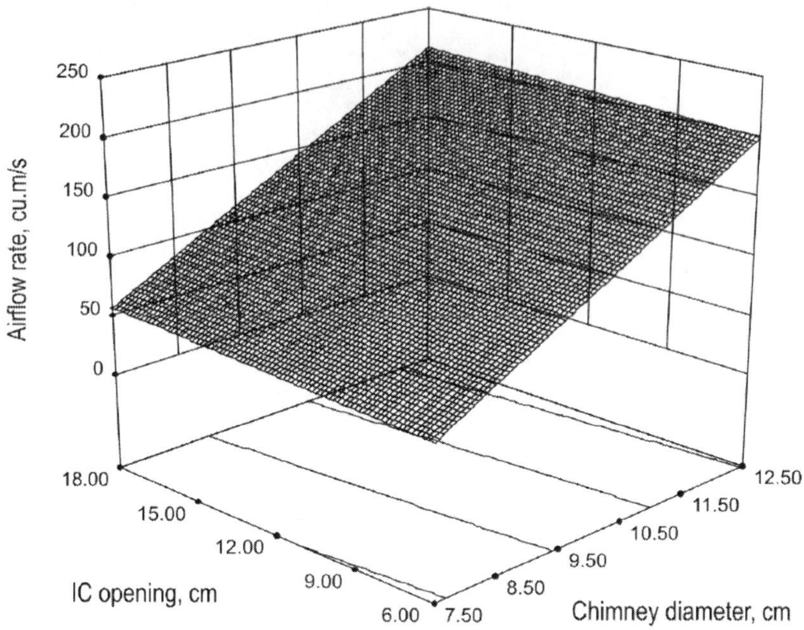

Figure 11. Response surface plot of fitted model for airflow as a function of chimney diameter and IC opening.

Temperature Results

The temperature data used in the generation of the model were averages of the five temperature readings recorded at 5 minute intervals, each from the top and bottom points inside the IC. Figure 12 shows the relationship between the average temperature and the two carboniser parameters as described by Equation 3. The model showed that as the chimney diameter is increased, temperature increased, although at a higher rate at smaller openings of the IC than at larger IC openings. Conversely, the effect of the size of the IC opening was more significant at smaller chimney diameters than at larger diameters (see Appendix 2). Highest temperatures were achieved at a 12.5 cm diameter chimney combined with an IC opening within the range of 12 to 15 cm.

$$\text{Temp} = -624.061 + 109.259 \times d_c + 58.878 \times h_{ic} - 2.343 \times d_c \times h_{ic} - 1.135 \times h_{ic}^2 \quad (3)$$

Oxygen Emission Results

The relationship between the O_2 content of the flue gas and the two carboniser parameters are shown in Figure 13, and can are described in Equation 4.

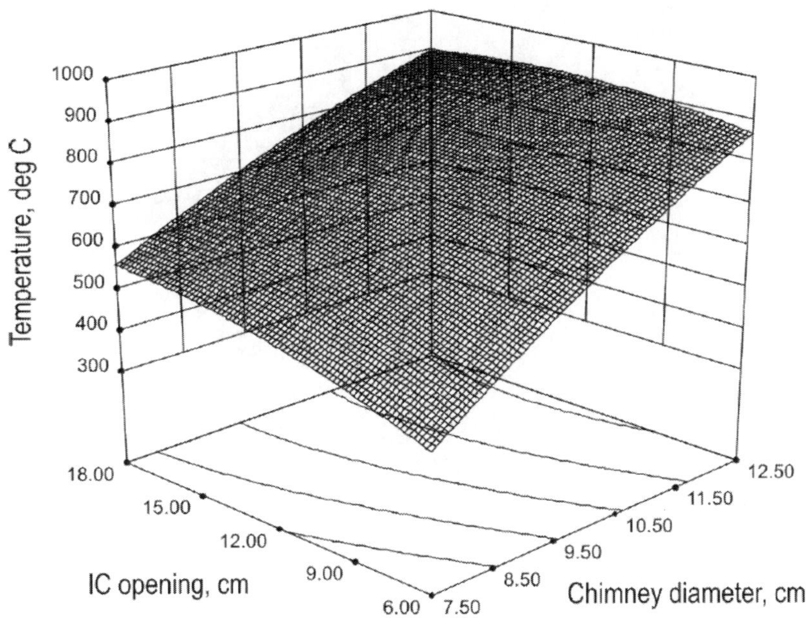

Figure 12. Response surface plot of fitted model for temperature as a function of chimney diameter and IC opening.

The presence of O_2 in the emitted gas from the carboniser showed that either or both the amount of air taking part in the carbonisation process is in excess of requirements, or the ambient air introduced into the IC was not properly mixed with the combustible gases and escaped through the chimney. The second possibility seemed to be the case when using the smallest chimney diameter (7.5 cm) in combination with the smallest IC opening (6 cm), as smoky emissions were observed alongside no visible flames in the IC. The highest O_2 concentrations were observed at the smallest chimney diameter and the rate of O_2 concentrations decreased at a higher rate as the diameter of the chimney was increased (see Appendix 3).

$$O_2 = -14.14506 + 8.40521 \times d_c - 0.16806 \times h_{ic} - 0.52393 \times d_c^2 \quad (4)$$

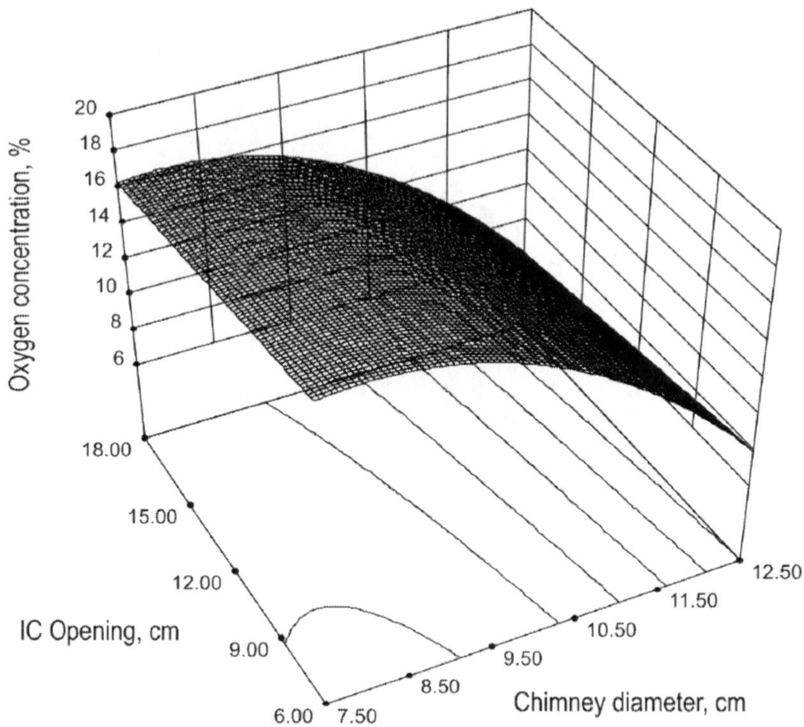

Figure 13. Response surface plot of fitted model for O_2 concentration in the flue gas as a function of chimney diameter and IC opening.

Carbon Monoxide Emission Results

Figure 14 shows the relationship between the CO content of the flue gas and the two carboniser parameters, which is shown mathematically in Equation 5. The effect of the chimney diameter is more significant than the size of the IC opening (see Appendix 4). The highest CO content could be observed at the smallest chimney diameter (7.5 cm) combined with a chamber opening of 10 cm, which decreases at an increasing rate as the chimney diameter is increased. The lowest CO content could be obtained at the largest chimney size of 12.5 cm regardless of the IC opening size.

$$CO = -453.975 + 418.793 \times d_c + 70.401 \times h_{ic} - 29.088 \times d_c^2 - 3.477 \times h_{ic}^2 \quad (5)$$

Carbon Dioxide Concentration Results

Figure 15 shows the relationship between the CO_2 content of the emitted gas and the two independent parameters, which, in mathematical form, is described in Equation 6. The CO_2 content of the flue gas indicates the presence or absence of combustion inside the IC.

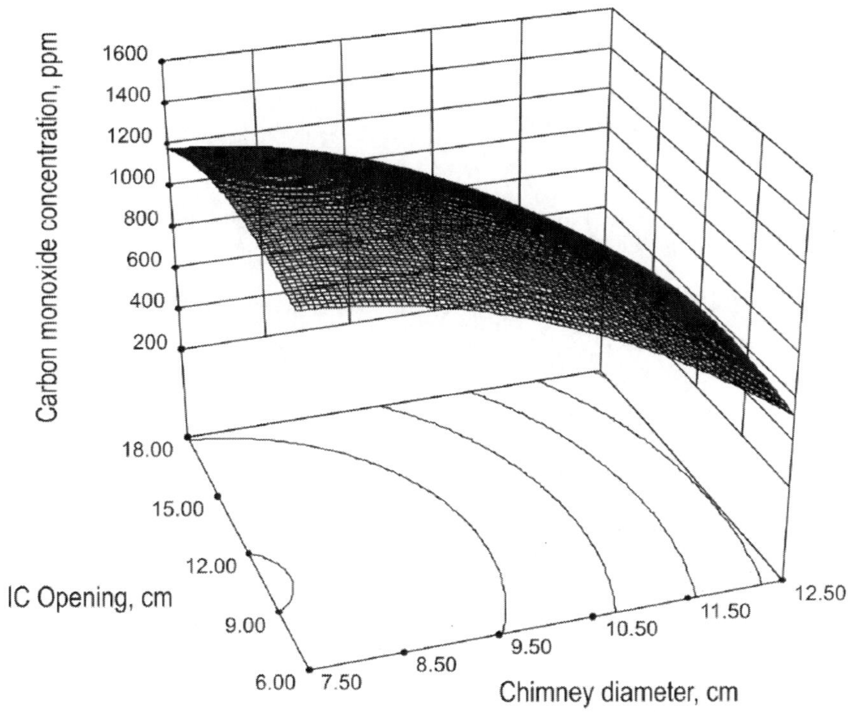

Figure 14. Response surface plot of fitted model for CO concentration in the flue gas as a function of chimney diameter and IC opening.

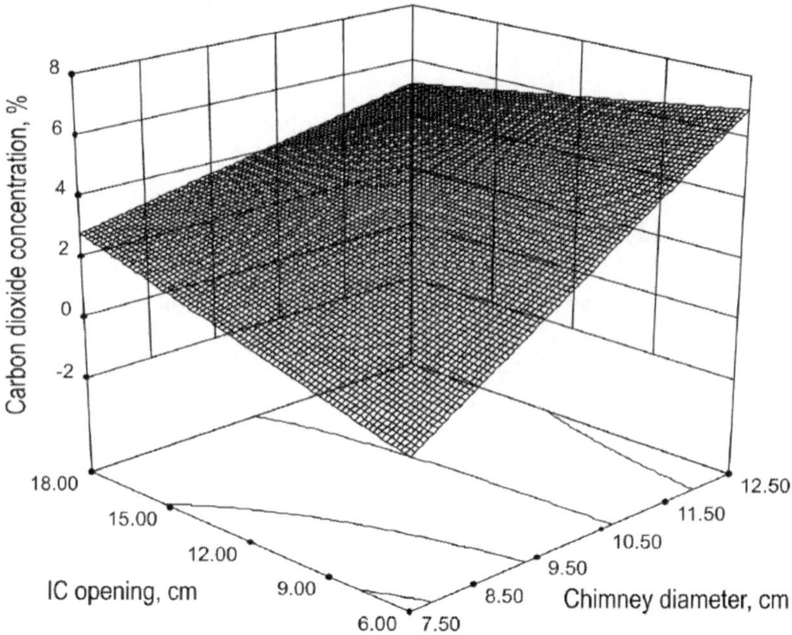

Figure 15. Response surface plot of fitted model for CO_2 concentration in the flue gas as a function of chimney diameter and IC opening.

Figure 15 shows that no CO_2 was recorded with the 7 cm chimney in combination with the smallest IC opening, and this was verified visually by no flames in the IC. Marked increases in CO_2 concentration were observed as the chimney diameter increased, especially at smaller openings of the IC, and the highest CO_2 concentrations were observed at the largest chimney and smallest IC opening combination. The effect of the chimney diameter on CO_2 concentration is highly significant, as shown in the analysis of variance (Appendix 5).

$$CO_2 = -17.03810 + 1.98667 \times d_c + 0.90833 \times h_{ic} - 0.084167 \times d_c \times h_{ic} \quad (6)$$

Biochar Quality Results

Using the rice husk biochar proximate analysis results, the analysis of variance (see Appendix 6) found no relationship between the fixed carbon content of the biochar and the carboniser parameters.

The model F-value of 2.39 implies that the model is not significant relative to the noise, and there is a 14.18 % chance that a model F-value is noise. Therefore, the independent variables had no significant influence on the quality of the biochar produced, and can be interpreted as optimised performance should be based solely on choosing independent variables that selected for temperature and emissions.

OPTIMISED COMBINATION MODELLING RESULTS AND CARBONISER TESTING RESULTS

Desirability functions were used by Design Expert® to optimise the carboniser performance emphasising the maximum temperature and minimum CO emission, see Equation 7.

$$D = \left(\prod_{i=1}^{n} d_i\right)^{\frac{1}{\sum r_i}} \tag{7}$$

where D = overall desirability considering all response factor in the optimisation, d_i = transformation of response variable i into a desirability function, r_i = is the importance factor for response variable i, and n = number of factors being optimised.

A desirability value of 1 (or close to 1) indicates an optimal condition. The numerical optimisation using Design Expert® yielded 23 combinations of the dependent variables (see Table 6 for the top five), with desirability values of at least 0.922. From the list, it appears that a chimney diameter of 12.5 cm is the optimum among the three sizes used. For the IC opening, Table 6 suggests that the carboniser can be operated within the range of 17 to 17.5 cm to attain high operating temperature and low CO emissions.

Testing was undertaken to verify the prototype carboniser performance when adjusted to the theoretical optimal combination of chimney size (12.5 cm) and IC opening (17.2 cm), and also to obtain a quantitive comparision with the conventional rice husk carbonisers in common use (see Table 7).

The tested performance results of the prototype continuous carboniser developed at PhilRice was superior to the conventional batch carboniser in terms of input capacity, maximum temperatures attained, and minimum CO emissions. Furthermore, in terms of biochar quality and yield, the prototype performed comparably with the conventional carboniser.

The final performance evaluation of the PhilRice continuous rice husk carboniser prototype demonstrated the conversion of 40 kg of rice husk per hour, at 40.3% biochar yields, with a purity of 99.1%. The prototype burned the gaseous products of carbonisation at the IC, prior to the chimney, reducing CO emissions to a low 431 ppm, with no visual smoke during operation, and achieved a maximum temperature of 899°C - a useful heat resource.

Table 6. Top five optimal combinations of the independent variables using Design Expert®

Number	Chimney diameter (cm)	IC Opening (cm)	Temperature (°C)	CO (ppm)	Desirability
1	12.49	17.29	913.2	417	1.000
2	12.50	17.19	914.8	419	1.000
3	12.49	16.78	917.7	441	1.000
4	12.48	17.22	912.8	425	1.000
5	12.48	17.57	910.0	404	1.000

Table 7. Performance comparison between the prototype analysed in this work, and the conventional rice husk carboniser design

Performance Parameter	Prototype	Conventional
Ignition time, min	3.2	2.4
Total amount of rice husk consumed, kg	164.5	99.5
Total operating time, h	4.1	4.08
Input capacity, kg rice husk/h	40.12	24.37
Char yield, %	40.3	40.8
Maximum temperature attained, °C	898.7	540.3
Attendance time, % of total operating time	21.3	14.6
Reloading with rice husk, %	8.6	-
Agitation, %	8.6	6.2
Collection of biochar, %	4.1	8.4
Purity of biochar output, %	99.1	99.2
Emission		
O_2, %	8.9	3.1
CO, ppm	431	1,405
CO_2, %	7.4	10.1

CONCLUSION

The current methods to covert rice husk residues in rural poor transitional economies are either open fires, or simple batch carbonisers, both of which expose rural workers to unacceptable levels of emissions [15]. This chapter describes a theoretical method to optimise small-scale bioenergy conversion technologies, suitable for improving the working environment of rural people in terms of air quality, and providing a means of capturing useful heat energy for additional renewable energy applications. Furthermore, any agronomic use of the biochar produced, while not a focus of this chapter, may improve the overall rice system primary productivity and reduce the level of imported nutrients required as inputs over time [16-19].

REFERENCES

[1] Jenkins, B. M. 1989. *Physical properties of biomass*. New York, US: Gordon and Breach; 1989.

[2] Marsh, H. and F. R. Reinoso. *Activated carbon*. Oxford, UK: Elsevier Ltd.; 2006. p. 479.

[3] Prakash, N. and T. Karunanithi. Advances in modeling and simulation of biomass pyrolysis. *Asian Journal of Scientific Research* 2009:2(1): 1-27.

[4] Bejan, A., G. Tsatsanoris and M. Moran. *Thermal design and optimization*. Canada: John Wiley and Sons; 1996.

[5] Jaluria, Y. *Design and Optimization of Thermal Systems*. New York: McGraw-Hill, US; 1998.

[6] Basu, P. *Combustion and gasification in fluidized beds*. Boca Raton, Florida, US: Taylor and Francis Group, LLC; 2006. p. 86.

[7] Ramirez, R. R., J. D. Martinez and S. L. Petro. Basic design of a fluidized bed gasifier for rice husk on a pilot scale. *Latin American Applied Research* 2007:37:299-306.

[8] Fang, M., L. Yang, G. Chen, Z. Shi, Z. Luo, and K. Cen. Experimental study on rice husk combustion in a circulating fluidized bed. *Fuel Processing Technology* 2004:85: 1273-1282.

[9] Lide, D. R. *CRC Handbook of Chemistry and Physics*. 88th ed. Boca Raton, Florida, US: CRC Press/Taylor and Francis Group, LLC; 2007.

[10] Denny, M. W. 1993. *Air and water: the biology and physics of life's media*. Princeton University Press, Princeton, New Jersey, US. p. 59.

[11] Joel, R. *Basic engineering thermodynamics*. 5th Edition. Edinburgh Gate, Harlow, Essex, England: Addison Wesley Longman Limited; 1996.

[12] Dean, A. and D. Voss. 1999. *Design and analysis of experiments*. New York, US: Springer-Verlag; 1999.

[13] Box, G. E. P. and K. B. Wilson. On the experimental attainment of the optimal conditions. *Journal of the Royal Statistical Society*. 1951:B13: 1-45.

[14] Pinheiro, E. R., I. M. D. A. Silva, L. V. Gonzaga, E. R. Amante, R. F. Teofilo, M. M. C. Ferreira, and R. D. M. C. Amboni. Optimization of extraction of high-ester pectin from passion fruit (*Passiflora edulis flavicarpa*) with citric acid by using response surface methodology. *Bioresource Technology* 2008:99:5561-5566.

[15] Orge, R., M. P. McHenry and R. de Leon. Modelling the carbonization process to develop a cost-effective, smokeless, continuous, down-draft rice husk carbonizer suitable for rice growing regions. In: A. Poullikka (Ed.), *Renewable energy: economics, emerging technologies and global practices*. Hauppauge, New York, US: Nova Science Publishers; 2013.

[16] Blackwell, P., G. Reithmuller and M. Collins. Biochar application to soil. In: J. Lehmann and S. Joseph (Eds.), *Biochar for environmental management: science and technology* (pp. 207-226). London, United Kingdom: Earthscan; 2009.

[17] McHenry, M. P. Agricultural bio-char production, renewable energy generation and farm carbon sequestration in WA: Certainty, uncertainty and risk. *Agriculture, Ecosystems and Environment,* 2009:129:1-7.

[18] McHenry, M. P. Soil organic carbon, biochar, and applicable research results for increasing farm productivity under Australian agricultural conditions. *Communications in Soil Science and Plant Analysis,* 2011:42:1187-1199.

[19] McHenry, M. P. Sensitive variables for applying biochar as a fertiliser substitute and a method to sequester carbon in soils: a wheat crop scenario. In: B. J. Ryan and D. E. Anderson (Eds.), *Carbon sequestration: technology, measurement technologies and environmental effects.* Hauppauge, New York, US: Nova Science Publishers; 2012.

APPENDICES

Appendix 1. ANOVA Table for the Airflow Response Surface Linear Model

Source	Sum of Squares	Df	Mean Square	F Value	p-value Prob > F	
Block	181.872	1	181.872			
Model	36986.945	2	18493.473	641.310	<0.0001	significant
A-Chimney dia.	36712.557	1	36712.557	1273.104	<0.0001	
B-IC opening	274.388	1	274.388	9.515	0.0115	
Residual	288.370	10	28.837			
Lack of Fit	252.173	6	42.029	4.644	0.0795	not significant
Pure Error	36.198	4	9.049			
Cor. Total	37457.188	13				
Std. Dev.	5.37			R-Squared		0.9923
Mean	129.61			Adj R-Squared		0.9907
C.V. %	4.14			Pred R-Squared		0.9907
PRESS	575.39			Adeq Precision		59.370

Appendix 2. ANOVA Table for the Temperature RS Quadratic Model

Source	Sum of Squares	Df	Mean Square	F Value	p-value Prob > F	
Block	44.14	1	44.14			
Model	2.732E+005	5	54646.23	68.42	<0.0001	significant
A-Chimney dia.	2.469E+005	1	2.469E+005	309.10	<0.0001	
B-IC opening	14528.76	1	14528.76	18.19	0.0037	
AB	4942.09	1	4942.09	6.19	0.0418	
A^2	1631.26	1	1631.26	2.04	0.1960	
B^2	2787.78	1	2787.78	3.49	0.1039	
Residual	5591.00	7	798.71			
Lack of Fit	3985.15	3	1328.38	3.31	0.1390	not significant
Pure Error	1605.85	4	401.46			
Cor. Total	2.789E+005	13				
Std. Dev.	30.05			R-Squared		0.9741

Appendix 2. (Continued)

Source	Sum of Squares	Df	Mean Square	F Value	p-value Prob > F	
Mean	712.89			Adj R-Squared		0.9611
C.V. %	4.21			Pred R-Squared		0.9327
PRESS	18755.60			Adeq Precision		26.217

Appendix 3. ANOVA Table for the O_2 RS Reduced Quadratic Model

Source	Sum of Squares	Df	Mean Square	F Value	p-value Prob > F	
Block	2.10	1	2.10			
Model	201.00	3	67.00	142.19	<0.0001	significant
A-Chimney dia	161.20	1	161.20	342.11	<0.0001	
B-IC opening	6.10	1	6.10	12.95	0.0058	
A^2	33.70	1	33.70	71.52	<0.0001	
Residual	4.24	9	0.47			
Lack of Fit	3.04	5	0.61	2.01	0.2585	not significant
Pure Error	1.21	4	0.30			
Cor Total.	207.34	13				
Std. Dev	0.69			R-Squared		0.9793
Mean	14.09			Adj R-Squared		0.9725
C.V. %	4.87			Pred R-Squared		0.9416
PRESS	12.00			Adeq Precision		30.187

Appendix 4. ANOVA Table for the CO RS Quadratic Model

Source	Sum of Squares	Df	Mean Square	F Value	p-value Prob > F	
Block	2619.45	1	2619.45			
Model	1.238E+006	5	2.476E+005	70.59	< 0.0001	significant
A-Chimney dia.	9.959E+005	1	9.959E+005	283.97	< 0.0001	
B-IC opening	36777.51	1	36777.51	10.49	0.0143	
AB	2.25	1	2.25	0.00064	0.9805	

Source	Sum of Squares	Df	Mean Square	F Value	p-value Prob > F	
A^2	90139.75	1	90139.75	25.70	0.0014	
B^2	42733.04	1	42733.04	12.18	0.0101	
Residual	24550.34	7	3507.19			
Lack of Fit	17770.72	3	5923.57	3.49	0.1291	not significant
Pure Error	6779.63	4	1694.91			
Cor. Total	1.265E+006	13				
Std. Dev.	55.40			R-Squared		0.9806
Mean	1037.71			Adj R-Squared		0.9708
C.V. %	5.34			Pred R-Squared		0.9136
PRESS	0.0000109			Adeq Precision		26.785

Appendix 5. ANOVA Table for CO_2 RS 2FI Model

Source	Sum of Squares	Df	Mean Square	F Value	p-value Prob > F	
Block	0.018	1	0.018			
Model	43.11	3	14.37	37.71	< 0.0001	significant
A-Chimney dia.	35.77	1	35.77	93.87	< 0.0001	
B- IC opening	0.96	1	0.96	2.52	0.1469	
AB	6.38	1	6.38	16.73	0.0027	
Residual	3.43	9	0.38			
Lack of Fit	2.43	5	0.49	1.94	0.2697	not significant
Pure Error	1.00	4	0.25			
Cor. Total	46.55	13				
Std. Dev.	3.33			R-Squared		0.3234
Mean	13.75			Adj R-Squared		0.1881
C.V. %	24.20			Pred R-Squared		-0.3348
PRESS	218.51			Pred R-Squared		4.3240

Appendix 6. ANOVA Table for the Biochar Fixed Carbon RS Linear Model

Source	Sum of Squares	Df	Mean Square	F Value	p-value Prob > F	
Block	7.29	1	7.29			
Model	52.95	2	26.47	2.39	0.1418	not significant
A-Chimney dia	52.21	1	52.21	4.71	0.0551	
B-Chamber open	0.74	1	0.74	0.066	0.8019	
Residual	110.76	10	11.08			
Lack of Fit	73.21	6	12.20	1.30	0.4175	not significant
Pure Error	37.55	4	9.39			
Cor. Total	170.99	13				
Std. Dev.	3.33			R-Squared		0.3234
Mean	13.75			Adj R-Squared		0.1881
C.V. %	24.20			Pred R-Squared		_0.3348
PRESS	218.51			Adeq Precision		4.3240

In: Charcoal
Editor: Eric N. Tyrone

ISBN: 978-1-62808-663-8

Chapter 3

Obtention of Charcoal from Eucalyptus Wood in a Steel Pilot Scale Kiln

N. Tancredi*[1,*2], *A. Cuña*[1,2], *J. P. Luizzi*[3], *M. Corengia*[1], *A. Sarachik*[1] and *A. Amaya*[1,2]

[1]Laboratorio de Fisicoquímica de Superficies,
Cátedra de Fisicoquímica,
DETEMA, Facultad de Química, Universidad de la República,
Montevideo, Uruguay
[2]Área de Energías Renovables, Polo Tecnológico de Pando,
Facultad de Química, Universidad de la República, Pando, Uruguay
[3]AFRECOR S.A. Valorización de Materiales Diversos.
Parque Tecnológico Industrial del Cerro, Montevideo, Uruguay

Abstract

The use of charcoal as an energy source has increased in developing countries in the last few decades. Charcoal production from wood has several advantages in regard to the direct use of forest biomass as fuel, among them: higher high heating value (HHV), lower moisture content and better resistance to microbiological degradation. Wood carbonization produces also condensable gases as tar, with its own value as fuel, and an

* Tel. +598 2924 8352, Fax +598 2924 1906, Email: nestor@fq.edu.uy.

aqueous phase (pyroligneous acid) that has found applications as a veterinarian product and herbicide. A lot of work has been produced recently about biomass carbonization. The results show the influence of temperature and heating rate on the process, as well as the char and condensate yields achievable. The effect of the lignin and cellulose content was also studied. Several kinds of kilns have been developed.

In this work, the obtention of charcoal and tar from sawmill residues (*Eucalyptus grandis* wood), for energetic uses was studied. A pilot scale steel kiln (capacity 100 kg) for wood carbonization and condensable gases recovery, based in that proposed by Unrinin and Yatagai was built. Some modifications were performed in order to improve the massic yields of charcoal and condensables, the energetic yield of the process, and the control of the smoke flow.

These modifications included the using of refractory bricks and ceramic fiber to increase thermal isolation, the addition of a chimney with a concentric heat exchanger as a condenser, a tar trap, a butterfly valve and a cyclone. The inner temperature at three different heights was measured and registered as a function of time. Several tests were carried out, varying the initial mass and monitoring temperature during ignition, pyrolysis, kiln cooling and unloading. Air inlet flow was strictly controlled. The whole process took about 20 h.

Raw material and products (charcoal, tar) were characterized by proximate analysis, HHV determination and thermogravimetric analysis in inert and air atmosphere, in order to study the influence of the carbonization conditions on the properties of the carbonization products.

The carbonization kiln produced a good quality charcoal (ash content and HHV in dry basis nearly 1 % and 30 MJ kg^{-1} respectively). Charcoal yields varied among 13 and 28 %. The energy conversion efficiency stayed between 34 and 47 %. Kiln modifications helped to overcome some troubles, such as incomplete wood carbonization. The recovery of condensate is possible, though only a low percentage of it is constituted by tar.

INTRODUCTION

Biomass is a renewable energy source whose use is increasing worldwide [1]. As Uruguayan National Policy tends to increase forest activity, biomass is becoming one of the most important renewable energy sources. In 2012, Uruguay had about 700.000 ha of artificial forests [2]. The forest products are mainly for the celullose industry, but also for energetic uses, especially as an alternative fuel in industrial steam generators or even as an exportation energetic product, shaped as chips. The national wood industry produces about

500 Mton of residues per year [3], with only part of them used in wood drying or electricity generation; the leftovers sometimes involve severe environmental damage. On the other hand, the national energetic situation is very oil-dependent and new government trends are searching for alternative solutions, such as the production and use of bioethanol and biodiesel for gasoline substitution. However, only firewood has been a solution for some industries. Wood industry wastes, such as wood sawmill and cuts are used sometimes, but its low density and heating value make very expensive its transport, therefore only part of them may be used in the generation place [4,5].

Charcoal production from wood has several advantages in regard to the direct use of forest biomass as fuel, among them major heating value and minor moisture content. Wood carbonization involves also the production of condensable gases as tar, that has also its own value as fuel or as a source of chemicals, and an aqueous phase (pyroligneous acid) that has found applications as a veterinarian product and fertilizer.

A lot of work has been produced recently about biomass carbonization. The results show the influence of temperature and heating rate on the process, as well as the char and condensate yields achievable [6-11]. Several kinds of kilns have been developed [6, 12, 13] and the effect of lignin and cellulose content was also studied [14].

The objective of this work was to study the obtention of charcoal and tar from wood residues, mainly for energetic uses. A pilot scale kiln was constructed for wood carbonization and condensable gases recovery, starting with the design proposed by Unrinin and Yatagai [15]. A series of tests were made to determine the massic yields of charcoal and condensables and the energetic yield of the process. Some modifications were made to improve charcoal and tar yields and the results were discussed. The raw materials (*Eucalyptus grandis* wood residues from a sawmill) and products (charcoal, tar) were analyzed in order to study the influence of the carbonization conditions on the properties of the carbonization products.

MATERIALS AND METHODS

Raw Materials

Sawmill residues (*Eucalyptus grandis* wood), were used in the carbonization tests. The samples were pieces of irregular form of massive wood, being the dimensions of the pieces of about 60cm x 20cm x 5cm.

Kiln

Carbonization tests were carried out in a kiln constructed for this purpose, based on the O-shaped design of Unrinin and Yatagai [15]. The kiln, with dimensions 130 cm (height), 180 cm (length) and 60 cm (width), and a capacity of 100 kg of wood, was made of sheets of steel. It had a removable door to load raw materials and unload charcoal; with two adjustable air inlets in order to control the combustion of the material inside the kiln. Thermal isolation was increased in comparison to the original design, by refractory bricks and ceramic fiber lining (Figure 1).

In order to improve the tar recovery yield and to have a better control of the smoke flow, a three inch chimney was installed at the back of the kiln, provided with a tar-trap.

Figure 1. Kiln interior and removable door.

Figure 2. Back of the kiln. From top to bottom: 1) chimney, 2) condenser, 3) butterfly valve crank, 4) tar trap.

A concentric pipe heat exchanger, with tap water flowing inside, which acted as a condenser, and a butterfly valve (Figure 2) were added in some of the tests. The valve avoided air entry through the chimney during the cooling stage. In an attempt to improve the condensate recovery without decreasing the chimney draught, in the last test the condenser was removed and a cyclone (scheme shown in Figure 3) was added instead. To be able to measure the temperature inside the kiln, three tubes were installed at the middle of the kiln length, at three different heights. Each one had one end blind and the other opened, in order to introduce external thermocouples. A fourth tube was placed at the chimney. In this work, only the central temperature is reported.

The tests involved: loading the kiln with different quantities of sawmill wastes (50 or 100 kg), ignition, control of air-inlet flow, temperature measuring, closing of every possible air entry once the desired temperature has been reached, monitoring of the kiln cooling and unloading of the kiln. Every complete test in the kiln (including cooling) took about 20 hours.

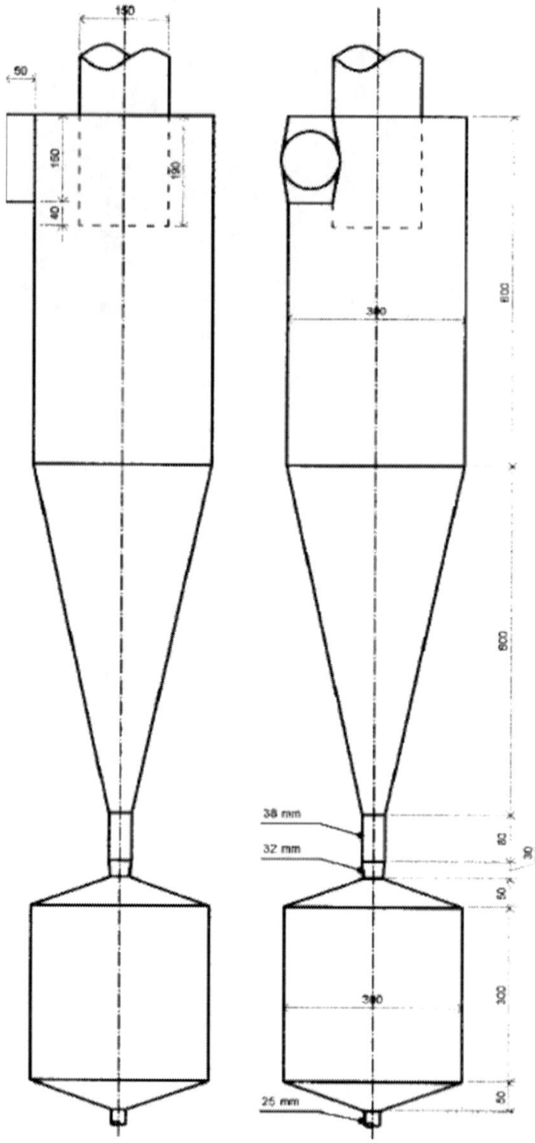

Figure 3. Cyclone scheme.

Analyses of Wood Wastes and Products

The products and the raw material were characterized by proximate analysis, determination of heating value and thermogravimetric analysis. An elemental analysis was performed on the raw materials and charcoal.

Proximate analysis included: moisture determination (ASTM D 2867-70), ashes determination (ASTM D 2866-70) and volatile matter determination (ISO 5621-1981). Water content for tar was determined by co-solvent (toluene) distillation.

Elemental analysis was carried out in a Carlo Erba model EA 1108 CHNS - OR equipment. For the determinations of heating value a Parr 1341 Plain Oxygen Bomb Calorimeter was used. Thermogravimetric analyses of the raw materials and products were carried out in the Shimadzu TG-50 equipment. The experiments were performed in inert atmosphere (N_2 99.995%, 50 cm^3min^{-1}STP, with a constant heating rate of 5°C min^{-1}, from ambient temperature up to 850°C) and in air atmosphere (50cm^3min^{-1}STP, with a constant heating rate of 2°C min^{-1}, from ambient temperature up to 850°C). Both gases were previously dried with a molecular sieve.

RESULTS AND DISCUSSION

Results for raw materials analysis are shown in Table 1.

Table 1. Proximate Analysis, Elemental Analysis and Heating Value of wood residues

Proximate Analysis	
Moisture (w.b., %)	10.5
Ashes (d.b., %)	0.16
Volatile matter (d.b., %)	87.8
Heating Values (d.b., MJ kg^{-1})	
High Heating Value	18.9
Low Heating Value	17.6
Elemental Analysis (d.b., %)	
C	48.4
H	5.9
N	0.0
S	0.0
O*	45.5

* calculated by difference.

Table 2. Conditions in the carbonization tests

Test	Operation	Wood mass (kg)	Maximum temperature (ºC)	Time to reach maximum temperature (h)	Cooling time (h)
1	Without condenser	58.60	520	3.7	19.1
2	Without condenser	100.46	636	4.5	13.9
3	Two-inches condenser	100.90	421	5.6	10.0
4	Two-inches condenser	49.93	510	3.5	12.4
5	One-inch condenser, internal grill	50.83	554	2.7	14.8
6	Cyclone, internal grill	53.03	503	1.7	10.0

Several carbonization tests were done in the kiln, with different quantities of wood and conditions of operation (Table 2). The maximum temperature indicated corresponded to the highest temperature measured in the central thermometer (this one was surrounded by the wood in all cases).

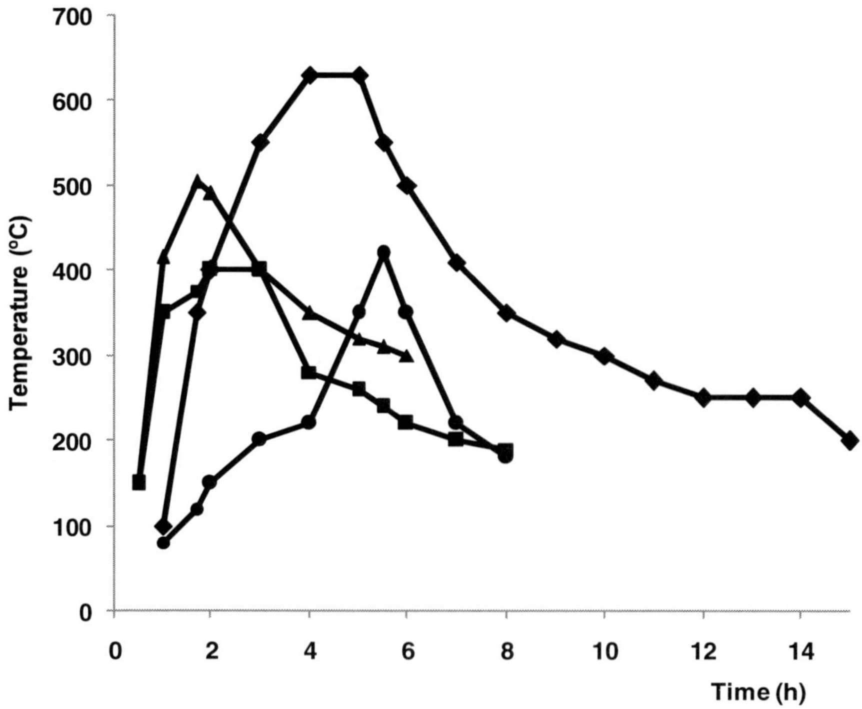

Figure 4. Temperature profiles of tests 2 (♦), 3 (●), 5 (■) and 6 (▲).

As a consequence of the different operation conditions the kiln performance changed, resulting in different temperature profiles and product yields. Charcoal and collected condensate fractions yields are shown in Table 3. The temperature profiles are shown in Figure 4 for a case without condenser (test 2), for a case with a two inch condenser (test 3), for the case with a one-inch condenser, and an internal grill (test 5) and for the case without a condenser and provided with the cyclone (test 6).

When the two inch condenser was added (test 3) the maximum temperature decreased and the time for reaching it increased, as a consequence of the reduction of the chimney draught. When the condenser was changed for a lower diameter one and a grill was included (test 5) the maximum temperature did not vary but the time to reach it was lower. In this case, these results are related to the improvement of the chimney draught. For the same reason, when the condenser was changed by the cyclone, the time to reach the maximum temperature was the lowest and the maximum temperature reached was of 500 ºC.

The values of some characteristic parameters for the tests are shown in Table 3. These parameters are defined as:

$$\% \text{ Charcoal Yield} = (\text{mass}_{\text{charcoal}} / \text{mass}_{\text{initial wood}}) \text{x} 100 \quad (1)$$

$$\% \text{ Collected Condensate Yield} = (\text{mass}_{\text{condensate}} / \text{mass}_{\text{initial wood}}) \text{x} 100 \quad (2)$$

$$\% \text{ Incomplete Carbonized Wood} = = (\text{mass}_{\text{not carbonized wood}} / \text{mass}_{\text{initial wood}}) \text{x} 100 \quad (3)$$

Taking into account that in some tests the carbonization was not complete, the previous parameters were corrected as follows:

$$\% \text{ Carbonization Yield} = (\text{mass}_{\text{charcoal}} / \text{mass}_{\text{converted wood}}) \text{x} 100 \quad 4)$$

$$\% \text{ Corrected Collected Condensate Yield} = = (\text{mass}_{\text{condensate}} / \text{mass}_{\text{converted wood}}) \text{x} 100 \quad (5)$$

The expected charcoal yield on a dry basis and for wood pyrolysis in this range of temperatures and at atmospheric pressure is about 37% [10]. The charcoal yield on a dry basis reached in test 2 (without condenser at the chimney) was 31.7 %, indicating a good kiln performance for charcoal production. Despite the variation of other conditions in the process, there was found an association between charcoal yield and maximum temperature attained (Tables 2 and 3).

Table 3. Tests results

Test	1	2	3	4	5	6
Charcoal Yield (%)	18.2	28.3	13.4	16.0	23.1	26.8
Collected Condensate Yield (%)	0	0	7.9	15.2*	6.5	7.6
Incomplete Carbonized Wood (%)	0	0	55.7	28.7	10.0	5.0
Carbonization Yield (%)	18.2	28.3	30.2	22.4	25.7	28.2
Corrected Collected Condensate Yield (%)	0	0	17.8	21.3	7.2	7.9

*mainly aqueous phase.

Table 4. Charcoal analysis

Charcoal	2	4	5	6
Moisture (w.b., %)	6.5	2.3	2.4	8.0
Ashes (d.b., %)	1.1	1.0	1.0	3.1
Volatile matter (d.b., %)	19.3	32.9	26.6	16.0
Fixed carbon (d.b., %)	79.6	66.0	72.4	80.9
High heating value (d.b., MJ kg^{-1})	29.6	28.0	29.7	29.6

The condensate is composed of an organic phase (tar) and an aqueous phase. Previous to the installation of the two inch diameter condenser pipe (test 3 to 5) the condensate fractions collected were not appreciable (tests 1 and 2). The chimney flow was affected by the installation of the condenser, because of the decreasing of the free internal area and the cooling of the gases; therefore from test 3 to 5 some chimney draught troubles had to be solved with the use of air forced flow.

Test 5 involved more changes; a grill was installed in order to improve the gas flow through the kiln, and the condenser diameter was changed from two inches to one inch to improve the chimney draught. This change brought an increase in the charcoal yield, but on the other hand, the condensate yield decreased.

In test 6 the use of the cyclone resulted in high charcoal and carbonization yields and a low percentage of incomplete carbonized wood. The condensate recovery, including tar and an aqueous phase, was slightly higher with regard to test 5.

Charcoal obtained from tests 2, 4, 5 and 6 were characterized by proximate analysis and determination of heating value (Table 4). Elemental analysis (Table 5) and thermogravimetric analyses (Figures 4 and 5) were carried out with the charcoal obtained in test 2, as it was the test with the best charcoal yield.

After test 3, tar in the trap was collected and analyzed (Table 6).

As expected, the charcoal produced in every test had better properties as fuel than wood samples: higher heating value, less moisture and less volatile matter (wich implies a higher fixed carbon content).

Table 5. Elemental analysis of charcoal produced in Test 2 (d.b., %)

C	H	N	S	O*
76.3	2.5	0.0	0.0	21.2

*calculated by difference.

Table 6. Tar analysis

Water content (w.b., %)	14.8
Ashes (d.b., %)	4.1
Volatile matter (d.b., %)	67.9
Fixed carbon (d.b., %)	28.0
High heating value (d.b., MJ kg-1)	21.6

In order to compare the results, taking into account the different amounts of incomplete carbonized wood and the low recovery of tar, a parameter defined by Antal Jr. and Grønli [10], the Energy Conversion Efficiency (ECE) was used:

$$\%ECE = (HHV_{charcoal} \text{ x } mass_{charcoal}/HHV_{converted\ wood} \text{ x } mass_{converted\ wood}) \text{x}100 \quad (6)$$

The results for some tests are shown in Table 7:

Table 7. Energy Conversion Efficiency

Test	2	4	5	6
Energy Conversion Efficiency (%)	47.0	34.4	41.7	44.1

Despite the fact that the higher %ECE and the lower percentage of incomplete carbonized wood was observed for test 2, in that assay no condensate was collected. From tests 4 to 6 tar was separated from the liquid phase in all cases; moreover the %ECE increased and the not carbonized wood decreased. Therefore, better results were obtained for test 6.

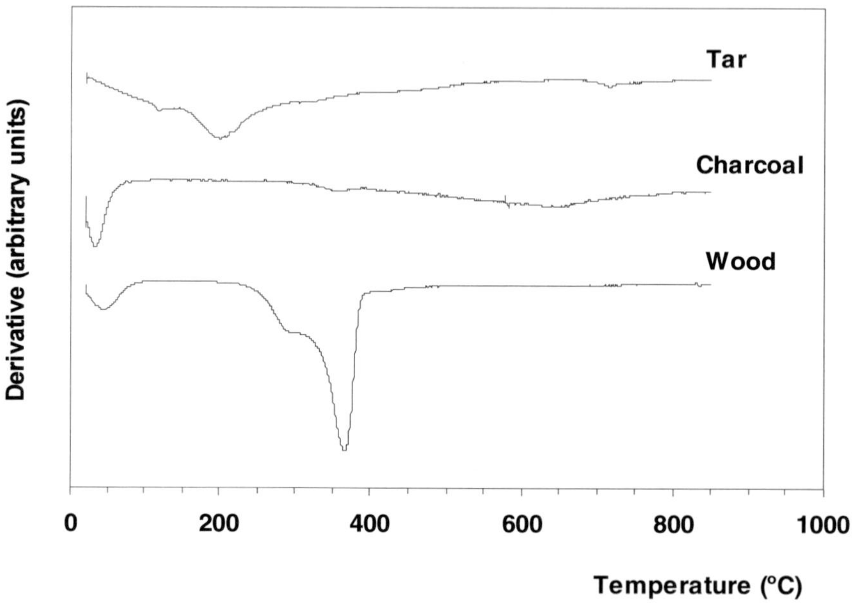

Figure 5. DTG in N_2 for wood, charcoal and tar.

Termogravimetric analysis was carried out in N_2 and air atmospheres for the wood residue, the charcoal obtained in test 2 and the tar collected after test 3. In N_2 atmosphere (Figure 5) the wood residue shows, after drying, a definite peak at 366 ºC, atributtable to lignin pyrolysis, with a shoulder centered at about 300ºC, due to cellulose and hemicellulose pyrolysis and in agreement with the high volatile matter content of the samples. Tar has a less important mass decrease and charcoal has almost no changes during heating, as a result of the devolatilization process during wood pyrolysis.

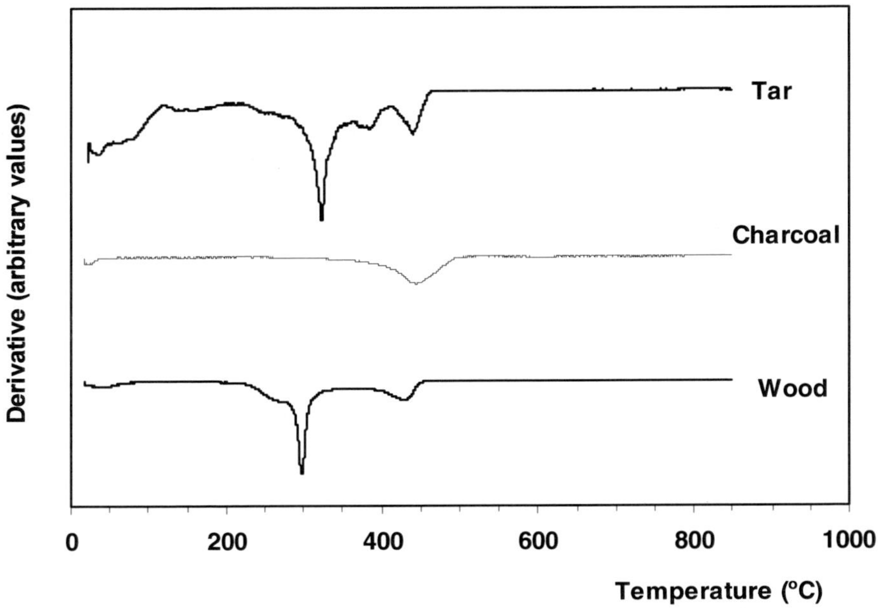

Figure 6. DTG in air for wood, charcoal and tar.

The DTG in air (Figure 6) shows an autoignition point of 250 °C for wood, 300 °C for tar and 400 °C for charcoal. These values indicate an increasing chemical resistance to combustion as a result of the thermal treatment and the composition of the products.

CONCLUSION

A steel kiln was used for the carbonization of wood sawmill residues. Different modifications introduced in the kiln lead to an improvement of charcoal yield, and a decrease of the percentage of non-carbonized wood. The carbonization kiln produced a good quality charcoal (low ash content and HHV nearly 30 MJ kg^{-1}). The kiln modifications introduced helped to overcome some troubles, such as incomplete wood carbonization. The recovery of condensate is possible, though only a low percentage of it is constituted by tar. From the energetic point of view, the carbonization in this kind of kiln may achieve good energetic conversion efficiencies.

ACKNOWLEDGMENTS

This work was financed by Project PDT 47/08, DINACYT, Ministerio de Educación y Cultura. The elemental analysis was performed at "Departamento Estrella Campos", Facultad de Química, Universidad de la República.

REFERENCES

[1] Pande, M. Bhaskarwar, A. N. In Biomass Conversion. The Interface of Biotechnology, *Chemistry and Materials Science*; Editors Baskar, C., Baskar, S., Dhillon, R. Springer: Berlin, Heidelberg, 2012, pp 1-90.

[2] Ministerio de Ganadería, Agricultura y Pesca. Internet portal of the Dirección General Forestal, www.mgap.gub.uy, acceded on May 2nd, 2013.

[3] Faroppa, C. Energy Consulting Services SA, Consultory to the Ministerio de Industria y Energía, 2010.

[4] Dirección Nacional de Ciencia y Tecnología. Noticias DINACYT, VIII (253); Ministerio de Educación y Cultura: Montevideo, March 17th, 2005.

[5] Dirección General Forestal, *Boletín Estadístico*, 4 (3); Ministerio de Ganadería, Agricultura y Pesca: Montevideo, December, 2004.

[6] Agirre, I.; Griessacher, T.; Rösler, G.; Antrekowitsch, *J. Fuel Process. Technol.* 2013, 106, 114–121.

[7] Elyounssi, K.; Collard, F. X.; Ngollo, J.; Blin, *J. Fuel* 2012, 96, 161–167.

[8] Tancredi, N.; Cuña, A.; Yoshida, M. In Pyrolysis: Types, Processes, and Industrial Sources and Products; Editors: Donahue, W.; Brandt, J.; Nova Science Inc. Publishers: New York, 2009, Chapter 8.

[9] Da Costa, P. C.; Sampaio, R. S.; Antunes, M. E.; Viana, E. A produção de carvão vegetal: teoria e prática; Author Edition: Belo Horizonte, 2006.

[10] Antal Jr., M. J.; Grønli, M. *Ind. Eng. Chem. Res.* 2003, 42, 1619-1640.

[11] Antal Jr,, M. J.; Mok, W. S. L. *Energ. Fuel.* 1990, 4, 221-225.

[12] FAO. Simple technologies for charcoal making; FAO: Roma, 1983.

[13] Muti Lin, *J. Fuel Process. Technol.* 2006, 87, 487-495.

[14] Gani, A.; Naruse, I. *Renew. Energ.* 2007, 32, 649-661.

[15] Unrinin, G.; Yatagai, M. *Jpn. Agr. Res.* Q, 1989, 23, 144-150.

In: Charcoal
Editor: Eric N. Tyrone
ISBN: 978-1-62808-663-8

Chapter 4

COMMENTARY: SIGNIFICANCE OF FUNCTIONALIZATION OF CARBON-BASED ADSORBENTS FOR WATER CLEANING TECHNOLOGIES

Margit Varga*[*]
Department of Analytical Chemistry, Institute of Chemistry,
L. Eötvös University

ABSTRACT

The characteristics of carbon can be varied according to the shape (granular, tubular, cloth etc.), porosity, specific surface etc. considering the requisites of technological applications. The properties of carbons are widely altered by the preparation method and by further functionalization and/or forming composites with other compounds (alumina, silica, or other minerals). The low price of raw material and carbon preparation technology is important in water cleaning technologies especially in developing countries. Further factors influencing the industrial use of the obtained activated (or modified) carbon are the environmentally friendly byproducts, end products as well as application processes. The carbon-based adsorbents saturated with hazardous inorganic and organic

[*] Corresponding author: Department of Analytical Chemistry, Institute of Chemistry, L. Eötvös University, P.O. Box 32, H-1117 Budapest, Hungary. Tel.: +36 13722500x1314; fax: +36 13722608. E-mail address: oltine@chem.elte.hu (M. Varga).

compounds should be reclaimed or recycled into the environment without any risk. The carbonization and activation method worked out in our laboratory permits the use of various raw materials resulting in the same end product. The properties of activated carbon can be easily designed by the functionalization of carbon-based low cost materials considering the above mentioned aspects. In order to remove various types of pollutants the chemical property of activated carbon adsorbent can be changed by varying the ratio of functional groups and carbon atoms fixed in carbonaceous matrix. This preparation method allows the combination of the carbonaceous matrix with other elements in order to obtain composites.

INTRODUCTION

Activated carbon is originally a crude form of graphite containing crystalline and amorphous structure alternating randomly. Due to its high porosity and a broad range of pore sizes it is capable to adsorb particles of various sizes (molecules, macromolecules, bacteria).

Therefore it is widely used in water cleaning technologies as well. Because it is constituted quantitatively mostly from carbon atoms the binding force of adsorption is characterized by week forces (Van der Waals force, namely hydrophobe-hydrophobe interaction) and the adsorption properties highly depend on the specific surface area as well as the pore size [1, 2, 3, 4].

One of the directions of carbon research is to increase the adsorption capacity by increasing the specific surface area, varying the pore size [5] and the shape of carbon particles (e.g. cloths [6]). This direction resulted in the elaboration of various activation methods [1, 2, 3, 4]. The activation can be performed physically (pyrolysis, microwave treatment [7]) and chemically.

My point is that in spite of the large number of activation methods the characteristics of activated carbon cannot be enhanced beyond a certain border by these ways. Moreover both of the physical and chemical methods increase the production expenses. By varying the shape it has the same effect as the activation since only the number, the availability and the size of the active sites can be changed. The production and application of carbon nanotubes seemed to be the breakthrough of this direction [8, 9, 10, 11]. The carbon nanotubes have extremely high specific surface and nano scale pore size which increase the number of active sites dramatically. The increase of specific surface and the decrease of pore size can result in the decline of adsorption speed (the reaction rate) because the reaction tends to be diffusion controlled.

The target analyte is forced to migrate outside and inside the nanotubes resulting in slower adsorption and the adsorbed analyte's latest desorption can be detained.

The other direction of the carbon research is the functionalization of the carbon atoms on the internal and/or external surface of the pores [2, 12, 13]. The functional groups are fixed by covalent bonds on the graphene surface and the carbon skeleton is only the holder of the functional groups. In this case the mesopores are advantageous because of their higher availability for the functionalization and for the adsorption of the target analyte during the application. The functionalization is performed by simple oxidation with strong oxidizing agents (e.g. concentrated acids [13]) resulting in carboxylic, lactonic, phenolic etc. groups or by the addition of a specific reactant to obtain special functional groups [2]. The functionalized carbon has another effect beside the physical adsorption. The functional groups bind the analyte by coordinative and/or ionic bonds. These bonds are stronger than the Van der Waals ones and might be both reversible and irreversible. In this case the characteristics of carbon adsorbent is determined by rather the functional groups than the carbon matrix therefore the characteristics can be easily fitted to the target analyte. The adsorption capacity depends on the number of functional groups rather than the specific surface area therefore it would be advantageous to increase it. Since the functionalization takes place on the surface and the dimension of the surface is limited, the surface cannot be extended without destroying the whole graphene skeleton.

Possible improvements of the carbon-based adsorbents can be the preparation of composites with other compounds [14, 15, 16]. In this case the matrix can be silica, alumina, or other minerals and the pores of the matrix are covered by carbon layers fixed by physical forces. Another possibility is the dispergation of carbon in a certain (e.g. polymer) matrix [17]. In the latter types of composites the components are present in a state of solid solution. In both cases the carbon can be functionalized. Besides of these methods the carbon atoms can be fixed on the surface with chemical forces increasing the stability of the composite. The advantages of the composites are the high specific surface area together with the increased number of functional groups.

In my opinion that during the design of a new carbon-based adsorbent the most important factors which should be taken into consideration are: costs of production, environmentally friendly preparation method, the fate of the used adsorbent in the environment as well as the reusability of the sorbent and/or the sorbate. The best way to decrease the production costs is to use waste materials or available cheap minerals for preparation of carbon-based sorbents.

Certainly these materials might not be effective enough therefore some modifications are needed. The modifications can be carbonization, activation and/or functionalization. Generally the environmentally friendly technology means avoiding the application of organic solvents, high energy consumption, dangerous side products etc. None of the available technologies can fulfill all of these conditions therefore the researchers must make compromises. For example the pyrolysis for carbonization needs high energy (e.g. 800-1000°C for 24 hours) and produces carbon dioxide and also the prepared carbon requires further treatment before application (e.g. activation with chemicals). The carbonization using concentrated acids produces volatile gases (e.g. SO_2, SO_3, NO_2) but needs moderate energy consumption (e.g. 100-200 °C for 1-2 hours) and does not need further activation. The production of carbon nanotubes requires the use of organic solvents and polymer matrices. In water cleaning applications essentially those technologies are the best for the production of carbon-based materials which could be performed by the simplest procedure, instruments and chemicals, the byproducts can be easily neutralized and which could be applied even in developing countries.

The next point is the fate and the effect of the produced new adsorbent and/or the end product of the application on the environment. The hazardous materials bind on the adsorbent by physical forces and can leach very easily. The most advantageous binding is when the adsorption/desorption of the target analyte depends on the pH. In this case the adsorption/desorption can be controlled by varying the pH. It seems to be optimal if the attached chemicals bind irreversible at around neutral pH-s (6-8) and can be reclaimed at lower pH-s in case of necessity.

It is essential to avoid the poisonous effects of the adsorbents which are synthesized for water cleaning. From this point of view the natural products or their simple modifications are advantageous. These adsorbents are environmentally friendly and do not cause serious hazardous effects for living organisms in case of their replacement to nature and can be submitted to bioregeneration [18]. Although the carbon nanotubes have several advantages in industrial applications, their use as adsorbent in water cleaning technologies would have inestimable effects for living organisms if they were produced and replaced into the nature in tons. The microorganisms are able to uptake the nanoscale object but they are not able to incorporate the carbon nanotubes. The carbon nanotubes destroy the membrane structure and kill the cell [9]. It has cytotoxic and genotoxic effect also for human health [20].

The research in the field of environmental chemistry needs complex knowledge and way of thinking so that the creation of even higher environmental pollution could be avoided. In our laboratory the research performed in the field of carbon science is based on the above mentioned principles.

SUMMARY OF OUR RECENT RESULTS

In our laboratory we apply chemical method for the carbonization using concentrated acids (sulfuric acid, nitric acid). This method is very cheap and does not require any special equipment in the technology. The developing gases are introduced through a container filled with CaO granulate and the reaction resulted in $CaSO_4$ and $Ca(NO_3)_2$ in air atmosphere. The applied temperature is 110-120 °C hence this is an energy saving procedure. This method does not require further activation and the number of functional groups can be designed by varying the ratios of acids and raw material.

The most important conclusions obtained from our results are the followings:

- The quality of the resulted charcoal did not depend on the used raw material significantly but it depended on the applied carbonization method. E.g. the carbonization by pyrolysis resulted in completely different charcoal characterization. The tested agricultural waste materials were cellulose based namely peanut shell, lapsi seeds and energy grass. There was not experienced any considerable difference among the charcoals obtained from the different raw materials in the maximal adsorption capacities, in the adsorption speed of the investigated target analytes and in other characteristics (e.g. FT-IR spectra, SEM images, surface functionalities determined by Boehm method, mathematical description of adsorption etc.). Therefore the method is easily adaptable for the application of locally available carbonaceous waste materials [20].
- The maximum adsorption capacities were in the highest range comparing to the present literary data for the tested hazardous compounds. The adsorption capacities strongly depended on the pH. In case of cations (e.g. Pb^{2+}, Ni^{2+}) and at the optimal pH values the adsorption was irreversible but shifting the pH to the lower direction desorption overwhelmed the adsorption and the adsorbed analyte could be reclaimed [21]. In case of anions [e.g. Cr(VI)] the pH dependency changed in the opposite direction comparing to the

cations and the adsorption was irreversible by increasing pH during the desorption experiments [20]. Applying As(III) in the same experiment there was no pH dependency in the investigated intervals [22]. The experimentally measured maximal adsorption capacities are summarized in Table 1 at the optimal pH-s using the same type of charcoals. Figure 1 shoes the experimentally measured and the calculated data applying the Langmuir model for Pb(II), Ni(II) and Cr(VI) and linear model for As(III). In the latter case the mathematical equation was y = 0.471 x (where y= adsorbed concentration in mg g^{-1}, x=initial concentration in mg L^{-1}, SD=0.018, R^2=0.986).

- According to the FT-IR spectra the heavy metal ions have bound to the carboxylic group and to the lactonic group by coordinative bonds which later transformed to ionic one during the drying [21] however the anions have bound to the phenolic and carboxylic groups. In this case the reactive forms are the protonated anions (H_3AsO_3 and H_2CrO_4). Based on the irreversible adsorption and the FT-IR results this bond is considered to be covalent [20, 22].
- The sorption isotherms could be modeled by the Langmuir and the Redlich-Peterson equation in case of lead(II), nickel(II) [21] and Cr(VI) [20] but using As(III) the isotherms were linear in the investigated interval at any pH-s [22].
- In case of the heavy metal ions the pH controlled desorption and respectively the irreversible bound in case of the anions allowed that the charcoals saturated with hazardous materials can be recycled into the environment without any risk of their leaching at natural pH-s.
- The thorough study of the charcoals treated with different concentrated acids revealed that the role of sulfuric acid was the production of double bonds between the carbon atoms by subtraction of water from the cellulose matrix. The role of nitric acid and the excess of sulfuric acid was the oxidation of the double bonds to carboxylic and lactonic groups [21]. The high excess of nitric acid could cause complete decarboxylation and decreased adsorption capacities since finally only the phenolic groups remained on the carbon matrix.

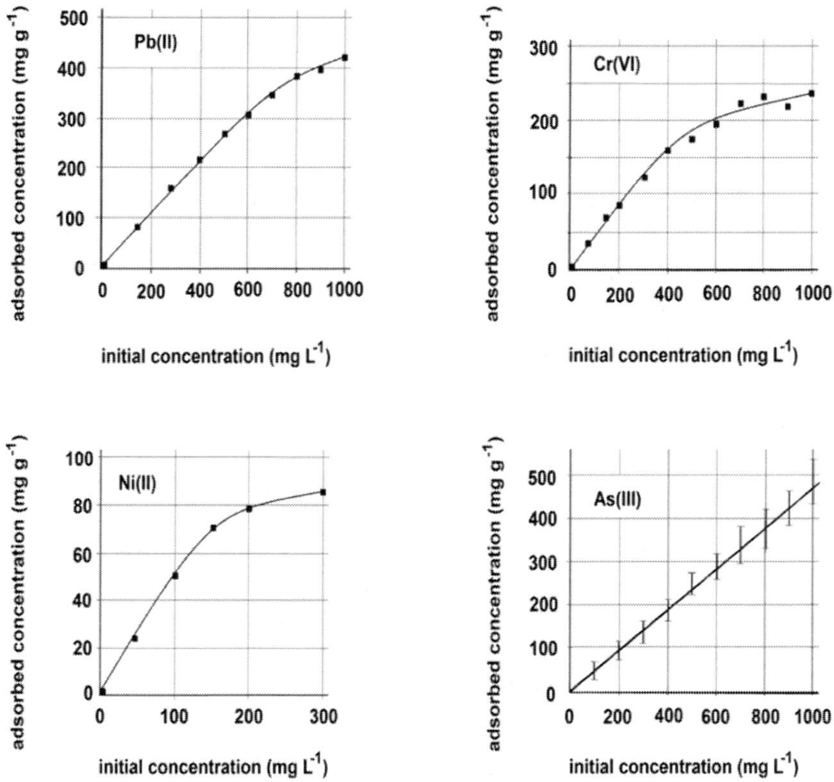

Figure 1. Experimentally measured (dots) and calculated data (lines) of various hazardous materials at the optimal pH-s according Table 1 (summarization of data in ref. [20, 21, 22]).

Table 1. Experimentally measured maximal adsorption capacities of various hazardous compounds at the optimal pH-s (summarization of data in ref. [20, 21, 22])

Compound	Optimal pH	Adsorption capacity, mg g^{-1} (RSD, %)	Adsorption capacity, mmol g^{-1}
Pb(II)	5	422.9 (±0.9)	2.04
Ni(II)	6	85.0 (±0.9)	1.45
Cr(VI)	1	230.5 (±4.4)	4.43
As(III)	1-9	469.8 (±8.6)*	6.27

* Average of the equilibrium values measured at nine different pH-s at 1000 mg L^{-1} initial concentration of As(III).

CONCLUSION

Referring to the Introduction part the functionalization cannot be increased without stint. In case of our method the funcionalization is limited by the oxidation of the double bonds. By the oxidation of one double bond one carboxylic and one lactonic group can be obtained [21] and in every oxidation step the quantity of carbon matrix decreases. At the end of this procedure the structure of carbon skeleton is destroyed and cannot hold the increased number of functional groups. With the combination of carbon with a suitable compound a composite matter can be produced and most probably the functionalization can be further increased. One of the directions of our scientific research is to prepare a highly functionalized carbon containing composite matter.

The composite is going to be prepared chemically using well characterized components having unambiguous composition so that the steps of the chemical procedure can be revealed. Furthermore the components should be substituted by waste materials for decreasing the production costs. Our further target is to apply the synthesized carbon composite for removal of hazardous materials and to extend our study for a wider range of heavy metal cations and anions as well as organic compounds.

REFERENCES

[1] Mohan, D; Pittman, CU. *J Hazard Mater B*, 2006, 137, 762–811.

[2] Yin, CY; Aroua, MKW; Daud, MAW. *Sep Purif Technol*, 2007, 52, 403–415.

[3] Lin, S-H; Juang, R-S. *J Environ Manage*, 2009, 90, 1336–1349.

[4] Dias, JM; Alvim-Ferraz, MCM; Almeida, MF; Rivera-Utrilla, J; Sanchez-Polo, M. *J Environ Manage*, 2007, 85, 833–846.

[5] Khalili, NR.; Campbell, M; Sandi, G; Golas, *J. Carbon*, 2000, 38, 1905–1915.

[6] Hoda, N; Bayram, E; Ayranci, E. *J Hazard Mater B*, 2006, 137, 344–351.

[7] Hesas, RH; Daud, WMAW; Sahua, JN; Arami-Niya, A. *J Anal Appl Pyrol*, 2013, 100, 1–11.

[8] Rao, GP; Lu, C; Su, F. *Sep Purif Technol*, 2007, 58, 224–231.

[9] Upadhyayula, VKK; Deng, S; Mitchell, MC.; Smith, GB. *Sci Total Environ*, 2009, 408, 1–13.

[10] Ren, X; Chen, C; Nagatsu, M; Wang, X. *Chem Eng J*, 2011, 170, 395–410.

[11] Pyrzynska, K; Bystrzejewski, M. *Colloid Surface A*, 2010, 362, 102–109.

[12] Mourao, PAM; Laginhas, C; Custodio, F; Nabais, JMV; Carrott, PJM; Carrott, MMLR. *Fuel Process Technol*, 2011, 92, 241–246.

[13] Zhao, N; Wei, N; Li, J; Qiao, Z; Cui, J ; He, F. *Chem Eng J*, 2005, 115, 133–138.

[14] Glover, TG; LeVan, MD. *Micropor Mesopor Mat*, 2009, 118, 21–27.

[15] Yang, GCC; Yen, C-H. *J Membrane Sci*, 2013, 425–426, 121–130.

[16] Shukla, PR; Wang, S; Ang, HM; Tade, MO. *Adv Powder Technol*, 2009, 20, 245–250.

[17] Ma, P-C; Siddiqui, NA; Marom, G; Kim, J-K. *Composites A*, 2010, 41, 1345–1367.

[18] Aktas, O; Cecen, F. *Int Biodeter Biodegr*, 2007,59, 257–272.

[19] Simate, GS; Iyuke, SE; Ndlovu, S; Heydenrych, M; Walubita, LF. *Environ Int*, 2012, 39, 38–49.

[20] Varga, M; Takács, M; Záray, G; Varga, I. *Microchem J*, 2013, 107, 25–30.

[21] [Shrestha, RM; Varga, I; Bajtai, J; Varga, M. *Microchem J*, 2013, 108, 224–232.

[22] Kauker, Z. Model examination of arsenic removal from natural waters, MSc Thesis, 1012, Institute of Chemistry, L. Eötvös University, Budapest, Hungary.

In: Charcoal
Editor: Eric N. Tyrone

ISBN: 978-1-62808-663-8

Chapter 5

CHARCOAL MEAL, AN OLD TEST BUT STILL VALID TO MEASURE INTESTINAL TRANSIT *IN VIVO*: EFFECT OF ACTIVATION OF THE TACHYKININ NK-2 RECEPTOR

Stefano Evangelista*
Menarini Ricerche SpA, Department of Preclinical Development, Firenze, Italy

ABSTRACT

Charcoal meal test is widely used for the measurement of gastrointestinal transit in small rodents. Here we described some results obtained with the activation of muscarinic, adrenergic and tachykinin NK 2 receptors on gastrointestinal propulsion in conscious rats. Charbachol, reserpine and the selective NK 2 receptor agonist, [βAla8]NKA-(4-10), produced a similar increase of the small intestinal transit in terms of 75-80 % travelled by charcoal as compared to values of 60 % recorded in controls. As expected atropine reverted the prokinetic effect of carbachol. The effect of [βAla8]NKA-(4-10) was restricted to the small intestine since gastric emptying was not affected by [βAla8]NKA-(4-10) and was atropine-resistant. Charcoal test is reliable and has been one of the first

* Correspondence to Dr. Stefano Evangelista, Preclinical Development Dept., Menarini Ricerche spa, Via Sette Santi 1, 50131 Firenze, Italy. Phone: +39/055/5680519; Fax: +39/055/5680419; e-mail: sevangelista@menarini-ricerche.it.

experimental model contributing to better understand the role of endogenous tackykinins in the gut motility.

Keywords: Charcoal, intestinal transit, neurokinin(NK)2 receptor, tachykinins

Tachykinins are a family of peptides that share the common C-terminal sequence Phe-X-Gly-Leu-Met NH2. In mammals, three peptides of this family, namely substance P (SP), neurokinin (NK)-A and NK-B, are believed to play a role as neurotransmitters, either in the peripheral or the central nervous system [1].

Tachykinins act as excitatory transmitters in the mammalian gut, where they exert potent spasmogenic effects via specific receptors, termed NK1, NK2 and NK3 and located on muscle cells or indirectly by activating intramural neurons [1-3]. "In vitro" studies showed that activation of NK2 receptor produces potent spasmogenic effects in several segments of intestine of various species [1].

To study this spasmogenic effect we have used activated charcoal that is carbon that has been processed to increase its porosity. It functions as an adsorbent; that is, atoms and molecules of certain substances will bind to its surface. Sources of medical-grade activated charcoal include wood and bamboo. It is usually supplied as a fine powder, either in capsules or in suspension. Activated charcoal is used in clinics to treat patients who have taken poison by mouth. The charcoal binds the toxins until they can be excreted, preventing them from being absorbed by the gastrointestinal tract.

Male Sprague-Dawley rats, weighing 225-250 g, were used. Food was withheld 24 hr before experiments but there was free access to drinking water. Recent investigations [4] showed that same results were obtained in rats and mice reducing the fasting period from a longer time up to 6 hr. Rats were given the selective NK2 receptor agonist, [βAla8]NKA-(4-10) (5-100 μg/kg i.p.), carbachol (10 μg/kg i.p.) or saline. In other experiments reserpine was administered at a dose (5 mg/kg i.p., 24 hr before) which produced an increase of intestinal transit comparable to the other stimulants. The meal, a 10 % charcoal suspension in 1 % carboxymethylcellulose, was given orally in a volume of 10 ml/kg immediately after the stimulant and the animals were killed 15 min later by cervical dislocation. The gastrointestinal tract was removed and the distance travelled by the marker was measured in the small intestine and expressed as percentage of the total length of the intestine from the pylorus to the cecum.

Gastric emptying was measured in 24 hr fasted male Sprague-Dawley rats, weighing 220-250 g. Rats were given [βAla8]NKA-(4-10) (15 μg/kg i.p.) or saline 5 min before 1.5 ml p.o. of a solution containing 1.5 % methylcellulose and phenol red (0.5 mg/ml). Fifteen min later the animals were killed by cervical dislocation.

In each experiments non-treated rats, receiving the phenol red solution, were sacrificed immediately thereafter to serve as standards (reference stomachs, 0 % emptying rate). Stomach were clampted at the pylorus and cardia, taken out and homogenized in 100 ml of 0.1 N NaOH. Proteins in 5 ml aliquots of homogenate were precipitated with 0.5 ml of trichloroacetic acid (20 % wt/vol) and centrifuged. The supernatant was mixed with 4 ml of 0.5 N NaOH and absorbance of the sample was read at a wavelength of 560 nm (A560) with a Perkin-Elmer spectrophotometer. Gastric emptying was calculated according to the following formula: gastric emptying % = (1- A560 sample / mean of A560 reference) x 100.

The following drugs were used: [βAla8]NKA-(4-10) (Peninsula), carbachol (Sigma), charcoal activated (Merck), phenol red (Fluka) and reserpine (Sigma).

Data were expressed as mean $\pm$ S.E. and the statistical analysis was performed by means of analysis of variance followed by Dunnett's test.

The activation of NK2 receptor through the selective agonist [βAla8]NKA-(4-10) produced an increase in intestinal transit which was significant at a dose of 15 μg/kg i.p. (73 $\pm$ 3 % vs 63 $\pm$ 3 % of controls; $P < 0.05$, n=8-10 for each group). Higher doses of [βAla8]NKA-(4-10) did not further increase the values of stimulated transit (72 $\pm$ 3 % for 50 μg/kg i.p. and 74 $\pm$ 5 % for 100 μg/kg i.p. of [βAla8]NKA-(4-10) vs 63 $\pm$ 3 % of controls; $P < 0.05$, n=8-10 for each group).

The increase of intestinal transit induced by [βAla8]NKA-(4-10) was of the same extent of that induced by carbachol or by reserpine acting trough the activation of muscarinic receptors and release of catecholamines, respectively. Reserpine was found to affect the gastrointestinal motility mainly via dopaminergic mechanisms [5].

Intestinal transit stimulated by [βAla8]NKA-(4-10) was unaffected by atropine at a dose which completely reverted the transit stimulated by carbachol (Table 1).

Table 1. Carbachol, [βAla8]NKA-(4-10) or reserpine induced increase of intestinal transit: effect of atropine

Treatment and Dose:	Intestinal Transit (%)
vehicle + saline	59 ± 2
vehicle + carbachol 15 μg/kg i.p.	75 ± 4 *
vehicle + reserpine 5 mg/kg i.p.	83 ± 3 **
vehicle + [βAla8]NKA-(4-10) 15 μg/kg i.p.	71 ± 2 *
atropine 100 μg/kg i.p. + carbachol 15 μg/kg i.p.	59 ± 3
atropine 100 μg/kg i.p. + [βAla8]NKA-(4-10) 15 μg/kg i.p.	68 ± 1 *

* and ** = $P < 0.05$ and 0.01 as compared to vehicle+saline group. n = 8-10 for each group.

The same dose of [βAla8]NKA-(4-10) used in previous experiments to stimulate intestinal transit was unable to produce a significant change in gastric emptying (57 ± 9 % for [βAla8]NKA-(4-10) (15 μg/kg i.p.) as compared to 56 ± 6 % for vehicle; n=8 for all groups).

The present findings show that activation of NK2 receptor by a selective agonist produces an increase in intestinal transit in conscious rats. This effect reached a plateau at a dose of 15 μg/kg since increasing the dose of the NK2 receptor agonist did not result in a further increase of small intestinal transit. The NK2 receptor-mediated increase in intestinal transit is restricted to the small intestine since gastric emptying was not affected by the same dose of NK2 receptor agonist. Conversely, other authors [6] previously showed that SP and NKA influenced gastrointestinal propulsion primarily by their effect on the stomach and pylorus. Methodological differences and the notion that NKA is not a selective NK2 receptor agonist since it possess a certain affinity for NK1 receptors could account for these discrepancies.

The observation that atropine was unable to affect this stimulated transit suggests that tachykinin NK2 receptor might play an important role in the non-cholinergic excitatory transmission to the small intestine. These data are in agreement with those of Auli et al [7] who showed that non-cholinergic contraction induced by nerve stimulation "in vitro" of human colon was mediated by activation of NK2 receptors. On the other hand, biochemical and autoradiographic studies have shown an abundant tachykinin-like immunoreactivity localized into nerve fibers of the muscular layers of rat and human small intestine and presence of NK2 receptors at this level [8-10]. Taken together, these pieces of evidence suggest that tachykinins released from nerve endings may mediate a potent spasmogenic effect in the small intestine. To this regard, it should be mentioned that increase of gastrointestinal motility induced by systemic administration of NKA can be recorded in healthy volunteers [11]. Interestingly, the action of NKA on gastrointestinal motility was accompanied by a number of adverse effects, part of which resembling symptoms characterizing the irritable bowel syndrome, like borborygmi, abdominal pain, headache, nausea, vomiting and flush (of the face or whole body) [11]. The involvement of NK2 receptors in the action of [βAla8]NKA-(4-10) was demonstrated by the selective blockade achieved with the polyciclic peptide NK2 receptor antagonist MEN 10,627 [12] which completely reverted the action of the NK2 receptor agonist without affecting the increase of intestinal transit produced by other stimulants such as carbachol and reserpine [13]. It was then confirmed by the use of a new generation NK2 receptor antagonist such as nepadutant [14]. It should be noted that tachykinins probably do not play a major role in intestinal motor activity under normal conditions, due to the modest/null ability of tachykinin receptor antagonists to modify normal peristalsis. This concept has been reinforced by studies performed in unanaesthetized dogs [15] and in healthy volunteers [11], in which the tachykinin NK2 receptor-selective antagonist nepadutant did not affect gastrointestinal normal motor activity per se. The use of reliable method such as charcoal test has contributed to allow the definition of effects induced by the release of endogenous tackykinins on specific targets in the gut and has let to develop a new class of drugs for controlling intestinal propulsive activity.

References

[1] Holzer, P. Tachykinins. In; Kastin AJ, editor. Handbook of biologically active peptides. 2nd ed, New York (USA): Academic Press; 2013, 1330-1337.

[2] Evangelista, S; Patachhini, R; Maggi, CA. Role of peripheral tachykinin receptors in neurogenic inflammation of the respiratory, genitourinary and gastrointestinal tract. *Curr Med Chem Anti-inflamm & anti-allergic agents*. 2003, 2, 157-174.

[3] Shimizu, Y; Matsuyama, H; Shiina, T; Takewaki, T; Furness, JB. Tachykinins and their functions in the gastrointestinal tract. *Cell Mol Life Sci* 2008, 65, 295-311.

[4] Prior, H; Ewart, L: Bright, J; Valentin JP. Refinement of the charcoal meal study by reduction of the fasting period. *Altern Lab Anim* 2012, 40, 99-107.

[5] Huang, W-J; Chien, EJ; Yeh, J-Y; Chen, J-J; Chiao, Y-C; Wang, PS; Chang, F-Y; Doong, M-L. The roles of dopamine receptor and adrenoceptor on the inhibition of gastric emptying and intestinal transit by amphetamine in male rats. *Chin J Physiol* 2012, 55, 259-266.

[6] Holzer, P. Stimulation and inhibition of gastrointestinal propulsion induced by substance P and substance K in the rat. *Br J Pharmacol* 1985, 86, 305-312.

[7] Auli, M; Martinez, E; Gallego, D; Opazo, A; Espin, F; Marti-Gallostra, M; Jimenez, M; Clavè, P. Effects of excitatory and inhibitory neurotransmission on motor patterns of human sigmoid colon in vitro. *Br J Pharmacol* 2008, 155, 1043-1055.

[8] Grady, EF; Baluk, P; Böhm, S; Gamp, PD; Wong, H; Payan, DG; Ansel, J; Portbury, AL; Furness, JB; McDonald, DM; Bunnett, NW. Characterization of antisera specific to NK1, NK2, and NK3 neurokinin receptors and their utilization to localize receptors in the rat gastrointestinal tract. *J Neurosci* 1996, 16, 6975-6986.

[9] Jaafari, N; Khomitch-Baud, A; Christen, MO; Julé, Y. Distribution pattern of tachykinin NK2 receptors in human colon: involvement in the regulation of intestinal motility. *J Comp Neurol* 2007, 503, 381-391.

[10] Nakamura, A; Tanaka, T; Imanishi, A; Kawamoto, M; Yoyoda, M; Mizojiri, G; Tsukimi, Y. Bidirectional regulation of human colonic smooth muscle contractility by tachykinin NK2 receptors. *J Pharmacol Sci* 2011, 117, 106-115.

[11] Lördal, M; Navalesi, G; Theodorsson, E; Maggi, CA; Hellström, PM. A novel tachykinin NK_2 receptor antagonist prevents motility-stimulating effects of neurokinin A in small intetsine. *Br J Pharmacol* 2001, 134, 215-223.

[12] Maggi, CA; Astolfi, M; Giuliani, S; Goso, C; Manzini, S; Meini, S; Patacchini, R; Pavone, V; Pedone, C; Quartara, L; Renzetti, AR; Giachetti, A. MEN 10,627, a novel polycyclic peptide antagonist of tachykinin NK2 receptors. *J Pharmacol Exp Ther* 1994, 271, 1489-1500.

[13] Tramontana, M; Maggi, CA; Evangelista, S. Spasmolytic effect of the NK2-receptor-selective antagonist MEN 10,627 in rat small intestine. *Jpn J Pharmacol* 1994, 65, 281-283.

[14] Tramontana, M; Evangelista, S; Giuliani, S; Manzini, S; Robelet, S; Girod, V; Maggi, CA. Influence of tachykinin NK2 receptors on intestinal sensitivity and motility in newborn rats. *Neuropeptides* 2010, 44, 269-272.

[15] Giuliani, S; Guelfi, M; Tolouse, M; Bueno, L; Lecci, A; Tramontana, M; Criscuoli, M; Maggi, CA. Effect of a tachykinin NK_2 receptor antagonist, nepadutant, on cardiovascular and gastrointestinal function in rats and dogs. *Eur J Pharmacol* 2001, 415, 61-71.

INDEX

A

B

C

D

E

F

G

H

I

J

K

L

M

N

O

P

R

S

T

U

V

W

Y

Z